INFORMATION SOURCES
FOR RESEARCH AND DEVELOPMENT

Use of
Physics Literature

INFORMATION SOURCES
FOR RESEARCH AND DEVELOPMENT

A series under the General Editorship of

R. T. Bottle, B.Sc., Ph.D., F.R.I.C., M.I.Inf.Sc.
and
D. J. Foskett, M.A., F.L.A.

Use of
Physics Literature

Editor
Herbert Coblans

BUTTERWORTHS
LONDON AND BOSTON
Sydney – Wellington – Durban – Toronto

THE BUTTERWORTH GROUP

ENGLAND
Butterworth & Co. (Publishers) Ltd.
London: 88 Kingsway, WC2B 6AB

AUSTRALIA
Butterworths Pty Ltd.
Sydney: 586 Pacific Highway, NSW 2067
Melbourne: 343 Little Collins Street, 3000
Brisbane: Commonwealth Bank Building, King George Square, 4000.

CANADA
Butterworth & Co. (Canada) Ltd.
Toronto: 2265 Midland Avenue,
Scarborough, Ontario, M1P 4S1

NEW ZEALAND
Butterworths of New Zealand Ltd.
Wellington: 26–28 Waring Taylor Street, 1

SOUTH AFRICA
Butterworth & Co. (South Africa) (Pty) Ltd.
Durban: 152–154 Gale Street

USA
Butterworths
161 Ash Street
Reading, Mass (Publishers) Inc. Boston, 01867

First published 1975

© Butterworth & Co. (Publishers) Ltd., 1975
ISBN 0 408 70709 7

Printed in England by Cox & Wyman Ltd.,
London, Fakenham and Reading

Preface

One of the difficulties in presenting the literature of any subject is the extent to which there is general agreement on its definition and scope. Traditionally, to the generation which learnt its physics in the 1920s a physicist was a physicist and we all used to know what that meant. Now, 50 years after, we are far less sure. On the one hand physics covers a much broader area with many multi-disciplinary fields and complex technical applications. On the other hand the physicist is inevitably mastering 'more and more' about 'less and less'. As one contributor (Chapter 8) puts it—'Theoretical physics, as a single subject, no longer exists, except perhaps as an occasional course-title in the undergraduate curriculum ... a physicist whose interest is primarily on the theoretical side will find himself working in a much more narrowly defined area, such as ... elementary particle theory.'

This may well turn out to be unfortunate for the health of 'natural philosophy' and could explain why there seems to be a shortage of those integrative masters for whom physics was essentially a unified whole. Be that as it may, the literature of physics can only partly be seen as identical with the component parts of physics itself. A bibliographical survey has to take a synoptic view and avoid too much fragmentation. This means using the older sub-divisions, since new and fashionable re-arrangements of sub-fields (e.g. solid state, condensed matter) take time to establish themselves internationally.

Thus some readers may well find that there are gaps and there is some duplication. However the aim has been to include a selection of the most useful printed sources. Also certain border fields will be found in other volumes of this series of information sources for R & D; e.g. biophysics in biology and geophysics in earth sciences. Admittedly the bibliographical style and the degree of selectivity are bound to vary in a volume that is a composite one with a number of authors, some of whom are working scientists while others are information specialists who work at another level. But in editing every effort has been made to ensure that the minimum elements for identification have been provided for each cited item.

The content and the form of this book had to be guided partly by the audience which it addresses. It is broadly intended for scientists, engineers, students, the new graduates and the various groups of information personnel. Of course it will be used by them for different purposes; the crystallographer will know the key texts of his own speciality, but may need information on the newer techniques of vacuum physics or the sort of physico-chemical detail that is easily accessible in one of the 'Handbücher'; the physicist in industry may have to move into a new industrial field, say, optical instruments for space research and thus need to refer to material that is well known to the academic. Therefore this survey has to be a compromise in which different users will approach the various chapters at different critical levels.

Finally I would like to express my indebtedness and thanks to the contributors, busy people for whom this work has been their contribution to our joint documentation responsibility; to INSPEC for supplying information and answering questions; to the Science Reference Library, and to the Aslib and CERN libraries for their help; to my wife for her deep involvement in matters linguistic and editorial.

H.C.

Editorial Note

An abbreviation is used in the text for publishers, particularly those that occur frequently. The full details of publisher and place are only provided once in an appendix.

Contributors

A. P. Banford, B.Sc., Head of Scientific Administration (at the time), Rutherford Laboratory, Chilton, Oxfordshire.

John W. Burchell, B.Sc., Science Reference Library, the British Library, London

Herbert Coblans, M.Sc., Ph.D., F.L.A., Managing Editor, Journal of Documentation, Aslib

Raymond H. de Vere, M.Sc., M.Inst.P., Science Reference Library, the British Library, London

Alison R. Dorling, B.A., A.L.A., Science Reference Library, the British Library, London

Peter D. Friend, B.Sc., Group Librarian and Information Officer, Atomic Weapons Research Establishment, Aldermaston, Reading

C.D.M. Johnston, M.A., C.Eng., M.I.E.E., Science Reference Library, the British Library, London

Elliot Leader, M.S., Ph.D., Prof. of Theoretical Physics, Westfield College, University of London

John A. Leigh, B.Sc., Science Reference Library, the British Library, London

Felix Liebesny, B.Sc., F.I.L., F.I.Inf.Sc., A.L.A., Transociates Ltd., London

Alan L. Mackay, M.A., Ph.D., Dept. of Crystallography, Birkbeck College, University of London

Elizabeth Marsh, Librarian, Rutherford Laboratory, Chilton, Oxfordshire.

A. Jack Meadows, M.A., M.Sc., D.Phil., Prof. of Astronomy and History of Science, University of Leicester

Ambar K. Mukherjee, M.Sc., Science Reference Library, the British Library, London

Jean G. O'Connor, B.Sc., M.Phil., University Library, University of Leicester

Brenda M. Rimmer, C.Eng., A.F.R.Ae.S., Science Reference Library, the British Library, London

John M. Ziman, F.R.S., M.A., M.Sc., D.Phil., Prof. of Theoretical Physics, University of Bristol

Contents

1

Introduction

J. M. Ziman

Where might one expect to find 'Physics' but in its literature? The knowledge of any one physicist is the minutest fraction of all that is written down, somewhere, recording the observations, opinions, conjectures, measurements of hundreds of thousands of people. The apparatus by which this information was obtained is merely an assembly of mechanical devices, of no significance out of the context of the knowledge to which they can be made to contribute.

What, indeed, is any science but a body of public, organised *knowledge*. Apart from the actual use of this knowledge in the affairs of men—which may be the prime employment of many readers of this book—the goal of scientific research is to collect information about the natural world, and to structure this information into a coherent, comprehensible form. The classical philosophical analysis of scientific method stops short, however, at the moment when the scientist, having arrived at some brilliant discovery, 'writes it up', for the benefit of an expectant world, and sits back waiting for the applause and bouquets. The truth is that new scientific information, however original and earthshaking, does not become scientific *knowledge* until it has been communicated, criticised, verified and remains unfalsified, by those sceptical vultures—other scientists in the same field. We have come now to realise the very great importance of the *communication system* of science as the instrument by which the results of individual private research projects are transformed into the *social* product of scientific knowledge.

The nature of this communication system is thus of the highest significance, especially for those who are professionally committed to basic research, or who have great need of particular types of scientific knowledge for technical application. The government bureaucrat without experience of research may imagine that a large,

1

expensive, but essentially automatic computerised index will do all that is needed: every graduate student, grappling with the background bibliography of his Ph.D. thesis topic, quickly learns differently. But not every high and mighty professor, with a hundred papers to his name, grasps the full subtlety and complexity of the system as it has evolved over 300 years.

The very title of this book—which specifies the *literature* of physics —indicates a restriction to the printed, published word, equation, or diagram. It is a familiar fact, however, that many things that 'every physicist knows' are never written down, but have been passed on, by word of mouth, from generation to generation. It is almost impossible to describe the precise manual procedures employed by a skilled craftsman: we learn them only by attempting to copy them under his personal supervision. Every good laboratory has its wise man, who can tell you the answer to a practical question that may never have been discussed at all in the official literature. In a complete account of the communication system we should also take into account all those little chats, private letters, seminars, conference talks, etc., which occur in the normal social contacts between scientists in the same discipline, and which play such a very great part in the early stages of the formulation of a new scientific idea. The substance of some of these informal verbal communications does often get into print eventually, but long after the moment when they were truly active upon their recipients. What Einstein wrote to Bohr in 1927 is of great historical, biographical and philosophical interest, but is not part of the current literature of physics in the narrow technical sense.

Nevertheless, the 'official' literature of physics is of perplexing diversity. A familiar error is to suppose that only the 'primary' literature counts—those innumerable, unintelligible, highly specialised accounts of particular little researches that are thrown together, quite without order, between the covers of the *'primary' journals*. To compound this error, it is widely held that only a very few leading journals really count—that if it has not appeared in the columns of the *Physical Review* it is not genuine physics, or can be disregarded for all practical purposes.

The fact is, however, that primary physics research is published in a large number of different journals, in many languages. It is quite true that various social processes tend to concentrate the work of the most active scientists in any given field into quite a small number—a few dozen, say—of internationally recognised journals, but this does not mean that a search for particular knowledge in the other journals is invariably useless. If we are not to go over to a monolithic, totally inflexible system of scientific publishing in which there is one and only one world-wide journal for each separate speciality, we must continue to insist that *it is the responsibility of every physicist to inform himself of the full literature of his subject, and to cite all relevant, reputable sources, at home and abroad.*

Notice, on the other hand, that there is very considerable specialisation of topics, even in primary journals claiming all of physics for their province. This *Hatton Garden effect* [district in London where the wholesale jewellers congregate] is a natural evolutionary process, which can be artificially accelerated by deliberate planning (for example, by the European Physical Society in its Europhysics journal scheme) but cannot easily be halted. The sub-division of some of the major journals of the big physical societies follows a similar trend. To 'know the literature' of a particular branch of physics is to know where one might expect to find the best new papers in that field, with perhaps an intuitive feeling for the scholarly style that is likely to be prevalent there, and a subtle grasp of the appropriate vehicle for some new paper that one would like to present to a particular circle of readers.

Prior to the primary journal in the public relations schedule of a new scientific discovery may come that bastard progeny of priority neurosis and reproductive technology—the *preprint*. Well, of course, it is very useful, occasionally, to know about some really good scientific idea a few months before it can actually be printed in hard type and distributed across the oceans by slow old mail boats, but this indiscriminate, *ad hoc* private distribution of badly printed versions of uncriticised scientific work evidently wastes a great deal of money, and dissipates the pressure that would otherwise build up to keep the official *public* journals running more efficiently. For my part, I don't count preprints as belonging to the literature of physics, and value 90% of those I receive chiefly as good scribbling paper!

The *letters journal* is more legitimate, but is often weakened by the attempt to perform two functions simultaneously. There is a need, indeed, for very rapid publication of preliminary brief reports of outstanding and essentially revolutionary new discoveries. There is also some need for a medium for the publication of quite short communications—not necessarily of great significance, but sufficiently self-contained and relevant to be useful to specialists in that field. Unfortunately, these functions are often confounded; most 'letters' in this type of journal are not complete in themselves, and yet do not promise such marvellous consequences that they deserve publication in advance of an ordinary primary paper in which all details would be made clear. But here, perhaps, we see most clearly the effects of the human pressure to get into print with original work as early as possible, without regard to personal reputation for clarity, accuracy or completeness of proof, or for our own needs as readers of all this stuff.

Even if we cannot expect all the scientific virtues from every scientific paper we are forced to read, we do insist on a certain level of scholarly accuracy and credibility. Indeed, a most extraordinary feature of all scientific research is the immense trust that we must put in the honesty and good faith of other scientists. We must simply assume, in the first instance, that when they say they have measured some property, or evaluated some formula, then this is literally true. Of course, after a little while, by critical comparison with our own results, or by contradiction with other equally sincere results quoted by other authors, we may come to doubt the accuracy of their work, and curse them for careless fools—but we know very well that reliable scientific results are very hard to achieve, and we take it for granted that they did their best.

This child-like trust in fallible, corruptible humanity is preserved by the *referee system*. When primary scientific journals were first invented, at the end of the seventeenth century, they carried a splendid miscellany of strange observations, factual or fictional, controlled only by the acumen or prejudiced judgement of the editor. But since these journals were usually published by learned societies, the privilege of publishing pretty much what they liked in them was restricted to the members of each society, or to their friends and clients who would 'communicate' through them. There still remain one or two journals of this kind—and there have been a few cases where a very distinguished scholar who has gone a bit gaga in his old age has given a very idiosyncratic twist to the publications in journals under his thumb. I am told, moreover, that the system of quality control in the Russian scientific literature depends more on the authority of the director of each author's institution than on the judgement of independent referees.

Generally speaking, however, a primary paper in a reputable physics journal has been subjected to scrutiny by one or more experts—almost always anonymous to the author—before being accepted for publication. The conflicts that can arise between authors and referees are legendary and searing to the spirit, but the experience of sitting on the editorial board of a journal and reading both sides will soon convince one that this system performs an essential service. This is the main bulwark against the publication of all manner of uncritical, irrational, essentially misleading and unsound work purporting to be science. Without some such explicit criticism, there would be no standards at all in research. The idea that good physics is, so to speak, guaranteed by the use of expensive equipment and complicated equations by people with Ph.D.s from accredited institutions is all nonsense. Research is an art that can be done very badly indeed, and the literature would be useless as a repository of scientific knowledge if it were not accepted that every paper that appears in a scholarly journal had been judged at least superficially sound by competent experts. The price to be paid when the experts mistakenly reject or delay an occasional good paper is negligible by comparison with the confusion, lack of confidence and general decline of scientific standards that would ensue if the anonymous referee system were simply dropped.

As I have indicated, the difficulty of doing good physics is so great that a high proportion (say 90%) of the papers that get published do not prove of permanent value. How does this process of purification and recrystallisation of the scientific literature actually occur? By what means are students, and others unfamiliar with a field over many years, guided to the choice of the truly significant works, containing the currently agreed knowledge of the topic in question? This is the function of the *review article*, the *monograph* and the *textbook*. The task of bringing together the results of primary research papers and presenting some unified account of progress in the solution of some particular scientific problem is of much greater importance and responsibility than is given credit for in the reward system of the scientific profession. It is recognised to be something of an art to do well, but by a shallow psychological assessment is accorded little originality by comparison with any petty application of conventional techniques to a 'new' primary research problem.

Of course, if a review is merely a descriptive bibliography of all papers on the subject, ordered by crude categories such as 'Experimental', 'Theoretical', 'Low Energy Experiments', 'High Energy Experiments', etc., then it demands little more than assiduity and a well-kept card index. This is the intellectual level at which such mechanical systems as abstract journals, information retrieval systems, and current awareness and alerting services can work with maximum efficiency. It must be emphasised, however, that despite their great utility as tools for *cataloguing* the primary literature of physics, and thus saving the immense labour of searching for potentially relevant material, these devices do not, and cannot by any stretch of the imagination, substitute for the intellectual work of making sense of what has been published, accentuating what is acceptable and rejecting what can no longer be considered worth attention.

There is a tendency to treat 'review articles' as a single category in the literature, and yet they are a genus with many species. Quite apart from specialisation by field—Nuclear Physics, Solid State Physics, etc.—they must be considered both as to audience and to aim. One journal, for example, may be an excellent medium for rather speculative reviews, very close to the primary literature and to current controversies, flying the kite of some particular theory over a somewhat wider range of potentially relevant evidence than would normally be allowed in an ordinary primary paper. Another type of review that is similarly addressed to the specialist research worker in a particular field might be a very careful compilation of experimental data, critically assessed and reduced to standard form for reference. Or it might be a formal account of a mathematical technique, bringing together and expounding implicit axioms, theorems and potentially useful formulae.

The next stage would be the specialist 'critical review', which would attempt to cover all the published literature on some particular problem over some particular period, indicating those points on which there is agreement, attempting to resolve contradictions of observation or theory, and setting out the elements of conflicting, controversial interpretations or explanations. Such a review is again usually intended for experts—and calls, of course, for considerable tact, generosity of spirit and intellectual firmness. If there is any meaning to the title of being an 'authority' on a subject, it must signify the ability to write such a review without losing either the respect or the friendship of one's scientific colleagues!

From the critical review, we move to the 'didactic' article, intended more for beginners in the subject, such as graduate students. Here the emphasis is on the clear exposition of what is well understood, in language intelligible to the relative outsider. This calls for special skills of style and thought, and the will to resist the blandishments of current controversies and speculations. It is particularly important to cultivate a tone of scepticism concerning many fashionable doctrines: this is the point where the unwary student may pick up paradigms that have not really been fully tested, and yet fail to grasp various simple basic points that every experienced worker in the field knows to be sound.

This is the level at which the coherent *monograph*, covering a much wider field than a single review article, can be extremely valuable. The greatest weakness of the communication system of physics is the absence of any mechanism for the deliberate production of syntheses of knowledge over any field broader than the conventional specialisations of individual research workers. Everybody will agree that such works are of the greatest value when they happen to get written, but there is no machinery for ensuring that appropriately qualified physicists actually embark on such projects and are appropriately rewarded at the end of them. Yet such a task is worthy of the very best efforts of any first-class scientist, over a period of years, and if carried out satisfactorily will make him a real authority over the whole field. It is true that symposia of separate reviews by independent authors are organised, collected together and printed as single volumes, under the guidance of energetic and learned editors, but these are usually lacking in intellectual coherence, and miss many valuable opportunities to make comparisons, draw parallels, carry over techniques from one problem to another, and generally to show that the world is not quite so complicated as the little bits of the jigsaw puzzle into which it is arbitrarily divided by scientific specialisation.

Presumably, this is the point where one should mention a class of literature that occupies quite a proportion of the shelf space in any physics library—the miscellaneous volumes of *conference proceedings* printed and sold to the great profit of many commercial publishing houses but bringing no great credit on those who conceive them scientifically. Superficially, it is an attractive idea to collect together and publish the papers presented by the galaxy of scientific stars that have gathered together for a week, from the four corners of the globe, on some delightful Aegean island, to scintillate to one another on, say, 'The Medium Low High Energy Electron–Proton–Meson Interactions in Galactic Clusters of Type 57b'. In practice, however, the 'contributed' papers are mostly shortened and inferior versions of work that is about to be, or has been published elsewhere, together with a few agreeable, racy, 'invited' papers which pretend to review the current situation but mainly convey the personal opinion of the speaker concerning the relative merits of his own and other people's contributions to the subject. Scientifically speaking, most volumes of this kind are almost worthless, and date rapidly, but they can, on occasion, be useful as the starting point for a proper literature search. If I really wanted to know about Electron–Proton–Meson Interactions . . . then I might glance through this conference report and note the existence of certain lines of argument, and go to the references cited in the various articles for a more accurate account. In other words, these are an expensive form of abstract service, suitable for the lazy man who has not kept up with all the literature.

The typical undergraduate or graduate *textbook* is a different animal again, organised around the peculiar needs of a course of study of so many weeks at such and such a level. It is characteristic, again, of the chaotic incompleteness of the literature of physics in the book format that there are innumerable excellent textbooks on certain standard courses—for example, undergraduate quantum mechanics or statistical mechanics—but almost nothing on new subjects that are only just entering the university curriculum. The commercial textbook publishers, by ruthless competition, are partly responsible for this unsatisfactory situation. Each publisher insists that he must offer a complete list of books on the conventional topics (not realising that academic physicists care nothing for an imprint, and only seek works by the best available author) instead of encouraging the publication of quite new books that will create their own fashion. And for any physicist who regards the writing of textbooks as mere drudgery, let me remind you of the historical fact recently unearthed by Gerald Holton—that Einstein got some of the essential elements of relativity theory from reading the perceptive textbook on electromagnetism by Föppl, who explicitly pointed out the anomalies in Maxwell's treatment of a conductor moving in a magnetic field. Textbooks eventually determine the scientific viewpoint of the generation who read them as students: there is as much opportunity for influence over the progress of science in this task as in many a great 'discovery'.

Scientific snobbery conventionally excludes from the 'literature' all works of popularisation. Yet journals such as *Scientific American* and *La Recherche* are immensely valuable, not only to the 'layman' but also to the professional scientist who wants a general picture of the present state of some distant branch of his own subject, or who would like to know, in simple terms, what all the fuss is about some fashionably famous discovery. The habit of reading such journals regularly is worth acquiring: it gives the physicist that continuing education which he cannot hope to get by more formal procedures. Every school and university science library should be well furnished with popular scientific books and magazines, to provide a background of general scientific culture, and a wider context for specialised courses of study. And every professional scientist should regard it as his duty to contribute to works of this kind, either directly, if he has the literary gifts, or through the intermediation of a skilled science journalist.

Indeed, the more one thinks about it, the more difficult it becomes to delimit the 'literature of physics'. In one direction we go into chemistry, in another into mathematics. Philosophy is not irrelevant to many of our deeper problems, and the practical connection with engineering is a broad artery along which valuable cargoes of information travel in both directions. Only a technical philistine would cut physics off from its own history; its place in modern civilisation; its cultural links; its religious, spiritual, aesthetic significance. If 'the literature of physics' means any book or journal from which a physicist might gain information that would profit him in his professional scientific activities, then it would surely not exclude the writings of Plato or the drawings of Leonardo!

But a literature is not merely the books on the shelves in some appropriately labelled bay of a library stack: it is a language, a style, a manner of talking. In some respects, the literature of physics is very homogeneous, with only a limited vocabulary and a restricted subject matter. It is well known that a large proportion of the technical literature of physics is written in a mixture of two international languages—bad English and algebra. Physicists of all nationalities are more concerned to put their ideas, however ill-expressed, to their colleagues in other nations than they are to use, with full elegance and subtlety, their mother-tongues. On the face of it, one would suppose that this would give a decisive advantage to those who have spoken some dialect of English from early childhood, but this is not evident from the literature itself. Perhaps the effort of expressing oneself in a foreign language clarifies thought; the worst scientific writing often comes from those who are so familiar with a language that they dash down the phrases as they come into their heads, stringing together cliché after cliché on the hope of muddling through to some final conclusion. There have been periods when scientific papers have been translated very clumsily into English by authors who have been cut off by war from close contact with that language, but this is not the present situation. There is no doubt, however, that insistence upon English as the *sole* language of physics, whether written or spoken, would impoverish our linguistic resources. There is more to be learnt from a sound professional translation of a paper conceived by its author in his own good Russian, Chinese or Spanish than from the conventional sentences in a standardised technical vocabulary in which he would be constrained to write if he were not really fluent in English. And it is for those of us who speak the various versions of English to respect this scientifically irrelevant factor, and to open our ears to foreign tongues and our minds to foreign thoughts.

It is true, nevertheless, that the subject matter of physics is so universal, so divorced from the varieties of individual human experiences, that it can best be expressed in a very narrow formal language where the words are deliberately stripped of their emotive significance. It is interesting to observe the evolution of this language. A vivid word used in a metaphorical sense to express a new concept— 'strangeness', 'spin-flip', 'tunnelling', 'saturation'—quickly takes its place amongst various more or less barbarous neologisms or acronyms—'superparamagnetism', 'laser'—in the jargon of our subject. In fact, the literature of physics is not so 'jargonised' as many other scientific disciplines—think of biochemistry, or geology—but a conscious effort is needed not to fall too deeply into an elaborate, stereotyped mode of expression. On the one hand, there is much need for a fairly precise vocabulary of words whose meaning can be clearly defined in a strict technical context; on the other hand, one should use a simple, clear, direct style, in words drawn from basic English or from more charmingly poetic sources, to impress the argument into the head of the reader. Such a style is not achieved without a real effort by the writer. It is often necessary to draft and redraft a paragraph a dozen times, listening to the words as if one had never heard them before, setting them in the most efficient logical order for the expression of the argument. The goal of scientific writing is not to initiate some outburst of passionate feeling in the reader but to persuade him that what you have to say could scarcely be doubted by any reasonable man. If you are confused, if you make grandiose claims, if you niggle away at irrelevancies while leaving great gaps in the argument, if you unconsciously contradict yourself—in other words, if he has any cause to suspect that you don't really know what you are talking about—then all your efforts are wasted. Your role is that of the friendly guide, taking the reader by the arm and gently leading him on, step by step, letting him see for himself the connections between the various points to which you will draw his attention, so that he feels, in the end, as much at home inside your argument as if he had lived there all his life. Physics is difficult enough to understand without the obfuscations of pretentious language, artificial elaborations of syntactical structure and similar devices used by vain simpletons to parade their superior learning.

The other major component of the literature of physics is *mathematical symbolism*. Now, of course, physics is deeply committed to the algebraic method of thought. I would, myself, define it as 'the science that attempts to describe the natural world in mathematical terms', emphasising not merely quantitative measurement but also the creation of strictly defined models whose theoretical properties mirror those observed in reality. The other major scientific disciplines define themselves by a natural realm of subject matter—the rocks of the earth's crust, say, or the mind of man—and exercise all forms of rational observation and discourse on that subject matter. In physics we have a special, immensely powerful intellectual technique, which we turn on nature in all its aspects but which only yields comprehensible results in particular circumstances. Physicists study atoms, electrons, nuclei, pure crystals, gases, etc., not because these are the topics allotted to us in the universal curriculum but because these are objects or systems that are actually simple enough to be described with fair accuracy by finite, soluble mathematical models. Mathematical symbolism is therefore an integral part of the literature of physics, and a physics paper without a single algebraic equation, integral, or table of numerical data can scarcely be imagined.

On the other hand, physics is not 'merely' mathematics. Perhaps one could express all the hypothetical properties of any of our theoretical models in the totally symbolic form of *Principia Mathematica*: there would still be the task of interpreting the symbols into the descriptive words of the world that they were supposed to represent. The programme of axiomatisation, being endless, turns out to be fruitless; at a certain stage we return to ordinary verbal propositions to express our discoveries and theories.

The choice between mathematical and verbal exposition is therefore, to some extent, a matter of taste. An equation may be intensely powerful and exact, but the symbols are cryptic, and do not speak so directly to the comprehension. It takes long practice and great concentration to 'read' a page of algebra and grasp its significance. To the extent, therefore, that the literature of physics consists of mathematical manipulations and formulae written out at full length, it becomes esoteric, and intelligible only after deliberate and laborious deciphering. The function of mathematical symbols is to allow one to carry out a sequence of complicated logical transformations and operations purely mechanically, without recourse to intuition. It is not very instructive, therefore, to set down in full all the successive stages of such a process; the important stages are at the beginning, when physical ideas are being represented by symbols according to some scheme of assumptions and approximations; and at the end, when the results of the calculation, numerical or graphical, are to be compared with experiment and interpreted. The experienced reader of physics does not try to follow through the mathematical argument, step by step, at a first reading, but jumps from verbal formulation of the first few equations to the discussion of the results, being capable of visualising for himself the general flow of the argument in between. It is the responsibility of the author of such a paper to bring into the open precisely these points of uncertainty of technique and interpretation, and to keep for his own notebooks and computer printouts the mechanical 'workings' of his calculations.

What is not, perhaps, adequately appreciated is the extent to which we actually depend on visual images rather than algebraic manipulation in thinking about physics. A deplorable tradition in applied mathematics—perhaps an excess of zeal by the followers of Descartes, in revolt against the geometrical proofs of Newton's *Principia* —was the deliberate banishment of diagrams from the exposition of physical theory. The literature of physics is now recovering from this impoverishment of the imagination, but there are still inhibitions, in the formal primary literature, against the pictorial representation of physical ideas. In this respect, the 'vulgarisers' have taught us a lot; faced with the difficult task of explaining very difficult ideas to the lay public, they have discovered, quite naturally, that a few lively diagrams speak far more eloquently than strings of words and symbols. There is no reason, except academic snobbery, why the same method should not be used to convey an intuitive notion of a new concept from the very moment of its birth.

The richness and variety of the literature of physics is confusing: the scale of it is daunting: the current journals on display in their hundreds, the stacks of bound volumes of journals, monographs, textbooks, data collections, annual reviews, etc. How can we find our way through such a mass of material to the few simple facts that we need to know? Who can read those 100 000 papers that are published each year? What does it all *mean*?

The great advantage that we have in physics is that the subject has a strong internal intellectual structure. Precisely because we seek a logical mathematical explanation of the phenomena that we study—indeed, confine ourselves to the study of phenomena that can be given explanations of this kind—we may discern in the physics literature a rational ordering of topics. It is evident, for example, that the study of the structure of the nucleus can be clearly distinguished from the theory of sound waves in liquids by the scale of the phenomena, in spatial extent, energy, etc. It is not too difficult, therefore, to set up a classification scheme, for the library shelves or for the index of a volume of abstracts, in which every topic seems to have its proper place.

Nevertheless, such schemes are often profoundly misleading. Nuclear structure is closely related to the dynamics of liquids by the 'liquid drop model'. The excitation spectra of atoms, of nuclei and of nucleons follow essentially the same quantum mechanical principles, although differing in energy scale by factors of 10^5. An alternative classification scheme, in terms of mathematical equivalence of models, would put all these phenomena on the same library shelf, at least within the same volume of an encyclopaedia of theoretical physics. In other words, the literature of physics is not like a biological 'family tree' but is multiply connected, and capable of classification in many orthogonal dimensions. For the librarian this may not be very important—he merely wants to find the proper niche for each book along some quasilinear scale. But for the reader, the searcher and, especially, the browser it is essential to know something of these connections, and to imagine the possibility that relevant information will be found in some quite different corner of the library, through a theoretical analogy, an instrumental similarity or the intervention of some mechanism on a completely different scale from those considered active. It is an astonishing fact about modern physics that the whole of our knowledge of the physical universe is interrelated: it is quite conceivable that our present uncertainties concerning *quasars*—the largest, most energetic, astronomical objects—may not be resolved until we understand about *quarks*—supposedly the smallest constituents of matter. To make full use of the literature of physics, one must really have an adequate grasp of this general structure, and not be constrained within the conventional boundaries of a subject as taught, of books as written, of libraries as arranged.

I have written almost as if the literature of physics were something external, public, to be found in libraries and consulted when necessary. But, of course, it is so much an integral part of the active life of a practising physicist that he must come to feel that he owns a bit of it himself. Books and journals are expensive to produce and to buy—but, to the extent that they contain our own products and our own needs, we must surely never grudge the cost. It is easy enough to estimate that the transfer of scientific information costs a few per cent of the whole of the research with which it deals—essentially a negligible proportion by comparison with its value to the scientific community. No serious physicist should, therefore, think twice before spending what is necessary for this purpose, whether by subscriptions to learned journals, the purchase of his own collection of books and reviews, or in support of a convenient library. Unfortunately, the system of grants, etc., which provide the material resources of research, does not always provide funds for such purposes, so that the 'literature' is treated as a marginal expenditure, coming from the taxed income of the individual research worker or from the tail-end of the university budget.

This is particularly the case in just those countries where the scientists are cut off, by distance and general poverty, from close personal contact with current research; nothing is more depressing, in a visit to a university physics department in a developing country, than to see the few meagre journals, often months late, on the library shelves, and to realise how weakly their attempts to contribute to scientific knowledge are sustained by a grasp of what has been and is being done by other people. But there is a 'parochialism of the heart', as well, to be found in scientific groups who have all the resources they need for such contacts, and who are yet too lazy, timid or foolishly vain to make themselves familiar with all relevant publications on their subject. It is often asserted that the literature is too vast, too incoherent, too overwhelming to be comprehended and used by the individual scientist, except perhaps with the aid of some mechanical, computerised, retrieval device. This does not seem to be the real truth. In an active research field the interests may be so narrow at each particular moment that only a few papers really count, and these can easily be discovered.

Let me repeat: it is the duty of the serious physicist to make himself aware of the literature of his subject—on the narrow front of his immediate research topic; more generally, on the broader front of his field; and in the large, as a man of genuine learning, over all of physics and the other sciences. Besides, it is all so *interesting*: are we not to be envied, in being permitted to devote our lives to such matters.

2

The literature of physics: its structure and control

Herbert Coblans

The general character of the literature of physics, like that of most other sciences, is determined by what might be called its ecology. This would include its historical evolution; its relationship with other fields, especially the exact sciences and philosophy; its structure in terms of the actual and potential needs of its users. The concepts of physics have always had a strong impact on both popular and academic thought: in our time relativity and quantum mechanics; and, on the other hand, the great clashes on science policy—above all, criteria for choice.[1] Are the enormous costs (relatively) of the high-energy accelerators to probe the structure of matter justifiable in relation to, say, research in molecular biology? Thus, it is not accidental that the juxtaposition 'little science, big science'[2] was formulated more precisely in the context of physics.

There is little doubt that an understanding of the mechanisms of information transfer in physics is important: obviously for the physicist, but also for the documentalists and the managers of science policy. This implies the need for more studies of how the literature of science grows and how it can be controlled bibliographically. It also means that the national organisation of physicists and the international arrangements for co-operation are becoming more and more relevant. This rather new but complex study of physics in all its aspects has been very well presented in a significant report published by the American National Academy of Sciences[3] under the title *Physics in perspective*, with a long chapter devoted to the information problems of physics.

In the first place, the dimensions of the literature, its structure and its dynamics, must be seen within the broader field of scientific information as a whole. While the identification of relevance for physics, in the somewhat chaotic book market, is not easy, the main difficulty is with periodicals and the proliferation in recent years of pseudo-publication in the form of technical reports, conference papers, translations and even preprints. The sheer weight of numbers is determinative here. Thus, INSPEC estimates that for 1973 the breakdown was:

Articles in periodicals	83·9%
Conference papers	14·7%
Books, theses, reports, etc.	2·4%

To this must be added the increasing importance of the literature of border disciplines, such as biophysics and electronics; of mission-oriented fields, such as nuclear energy and space research.

A study of the vital statistics[3,4] provides some indication of the nature of the problem, in spite of the fact that bibliometric data of this kind can be misleading, as the basis for counting and comparison is not always compatible. Also, as much of statistical work reported is Anglo-Saxon in origin, there is admittedly a bias towards material in the English language. None the less, the general picture which emerges is not likely to be seriously wrong. Thus, of the four truly comprehensive services* providing abstracts for the literature of physics, *Physics Abstracts* will be taken as the basis. (Cooper and Terry[5] describe 69 secondary services, but most of them are only supplementary in function. See Chapter 5 for more detail.)

ARTICLES RELEVANT FOR PHYSICS AND THE PERIODICALS IN WHICH THEY APPEAR

The periodicals strictly concerned with physics were around 800 in number in 1968 (there were 200 in 1920). The rate of growth of the number of titles has been roughly linear, on the average 12 per annum. (Currently *Physics Abstracts* obtains its input by scanning some 1600 periodicals.) On the other hand, the number of abstracts in *PA* has grown exponentially, doubling approximately every 8 years. This rate of growth is strikingly paralleled by that of the membership of the 25 member societies of the American Institute of Physics[6] and the total number of pages published annually in its own and translated periodicals (in both cases an 8 year figure for doubling). What is so misleadingly called the 'information explosion' is not an explosion at all, but rather the simple and logical consequence, more

* *Physics Abstracts* (London), *Physikalische Berichte* (Braunschweig), *Bulletin Signalétique* (Paris), *Referativny Zhurnal-Fizika* (Moscow).

physicists produce more papers.[7] A selection of the absolute figures for abstracts in *PA* follows:[4a]

1921	2 010	1968	50 480
1931	4 365	1970	79 830
1941	2 735	1971	84 332
1951	9 200	1972	85 185
1960	21 405		

This gives some indication of the load and the cumulating totals which must be handled for retrospective searching.

Adequate coverage is a matter of real concern to physicists and information officers. This is largely determined by the degree of scatter of the articles of physics interest in the whole periodical literature of science and technology. The trend of the scatter based on *PA* is shown in *Table 2.1*.

Table 2.1

	Total number of articles from periodicals	Yield of abstracts
1. 1964	29 000	90% from 124 periodicals
		60% from 34 periodicals
2. 1973	68 300	75% from 150 periodicals
		57% from 75 periodicals

1. Figures quoted by Anthony, East and Slater.[4a]
2. Figures supplied by INSPEC.

It is clear that in the past decade the scatter of the literature of physics has grown and the core of periodicals which could supply a reasonable coverage, say 90%, has become much larger.,

Another factor which is important for the user is the language distribution in the periodical literature. Here the English-speaking physicist has a great advantage, since 68% of the entries in *PA* have English as the original language. The proportions for the other languages are approximately (reference 3, p. 923):

English	68%
Russian	17%
German	7%
French	6%
all others	<3%

Cutting across and in a sense in conflict with the formal printed publications is the 'preprint', especially rife in high-energy and theoretical physics. It has become a method of direct personal communication rather comparable with the letters written to each other by natural philosophers before the growth of periodicals. Since they are not usually accepted as part of the bibliographical record, they have no public validity and tend to circulate within an 'invisible college'. A study carried out for the AIP[8] among high-energy theorists showed that on the average each member of the group received some 186 preprints per year. Although this practice offends against the universal nature of scientific communication, it usually gives the recipient at least an advance of 6 months on the appearance of the article in print (see below, p. 100). So far, apart from 'Preprints in particles and fields and antipreprints' (see p. 216), various schemes for the central organisation of the large-scale distribution of preprints have not been successfully launched.

In recent years many studies have been carried out on the use of different information channels for scientific communication. In the USA the general pattern seems to be that personal contact is at least twice as important as abstracts and indexes, that informal methods (preprints, reprints, bulletins of Information Exchange Groups) are used 50% more in practice for current awareness than the more formal methods. Thus, quite apart from their publications conferences play a significant part in fostering awareness and catalysing the flow of ideas. However, there is an undesirable tendency to underrate the importance and value of the published record of physics—what is sometimes disparagingly called the 'archival' literature.

Physics in our time, like the other natural sciences, is very much influenced by its international organisation. The International Union of Pure and Applied Physics (IUPAP) was created in 1922, when a strong movement of internationalism was seen as a necessary step towards rehabilitation. (Similar Unions were started for astronomy, chemistry, radio sciences and biological sciences.) This was the beginning of a realisation of the great importance of standardisation. Thus, the aims of IUPAP are, among others, to obtain international agreement on constants, notation, terminology and standards; to co-ordinate the work of preparing and publishing abstracts, papers and tables of physical constants; and to encourage international co-operation in physics.

Membership is by country, through its national academy, research council or government: a world federation of official national representation which can act as a mouthpiece for a given scientific discipline—in this case physics. In this way a formal structure is given to the international character which is inherent in scientific work. The work of the Union is mainly carried on by Commissions and the

General Assembly, which usually meets every 3 years. Among the 16 Commissions the most important from the point of view of information are 'Symbols, Units, Nomenclature (SUN)' and 'Publications'.

INTERNATIONAL COUNCIL OF SCIENTIFIC UNIONS (ICSU)

The full 'political' effectiveness of the individual Unions is achieved through their federation in one of the most influential of all international non-governmental organisations, ICSU. Founded as the International Research Council in 1919 it changed to its present form in 1931, and now includes 17 international Unions. Thus, it can be considered both a Parliament and a Cabinet for world science and the working scientists at the horizontal level of the distinct disciplines. Understandably, World War II greatly disrupted its organisation and it was one of the first NGOs to be rehabilitated by Unesco in the late 1940s. It still receives considerable sums from Unesco, e.g. $530 000 from the Regular Programme for the biennium 1973/74.

Its main objectives are to co-ordinate and facilitate the activities of the Unions, and to act as a clearing house for the countries (now over 60) adhering to the Council. Apart from its annual General Assemblies, part of its work is done through Committees devoted to broad international themes, e.g. the Committee on Space Research (COSPAR), the Committee on Science and Technology in Developing Countries (COSTED) and more recently the Scientific Committee on Problems of the Environment (SCOPE). The International Geophysical Year (IGY) in the late 1950s was typical of the sort of organisation which only the prestige of such a body as ICSU could carry through so successfully. It has resulted in a vast international network of World Data Centres which are a continuing source of basic information.

The Committee on Data for Science and Technology (CODATA) which was formed in 1966 has a special relevance for physics information. It aims to co-ordinate world-wide activities in the collection and evaluation of numerical data on the physical and chemical properties of substances and materials. This is done by conferences at approximately 2 yearly intervals, which bring together compilers from data centres, representatives from national academies, publishers, etc. There are Task Groups on 'Computer use', 'Key values for thermodynamics', 'Fundamental constants', 'Data for chemical kinetics', 'Presentation of data in the primary literature' and

'Accessibility and dissemination of data'. The Executive Director with his staff of experts has produced a very useful survey, the first authoritative one of its kind:

CODATA. *International compendium of numerical data projects.* Springer, 1969.

It describes more than 150 data centres and projects in 26 countries, listing and analysing their publications, and includes an account of the main international bodies for standardising units, symbols, constants and terminology.

The ICSU Abstracting Board was formed in 1953 as a result of a recommendation by the Unesco International Conference on Science Abstracting; the hope was that scientists would be drawn into more active participation in the secondary services. It started off with some disadvantages: the large traditional services for abstracting were somewhat exclusive in terms of their vested interests (their financial difficulties only became serious some years later); and the 'pure science' outlook of ICSU was not very meaningful in the context of a literature in which 'pure' and 'applied' are often hard to disentangle. Therefore, it started off slowly in physics and chemistry, later moving to biology. However, in recent years the climate has changed for abstracting services and the Board now has 16 Member Services, thus also bringing in mathematics, astronomy, geology, engineering and medicine. It has become an important clearing house for information questions, particularly by regularly bringing together the executives of the bodies which determine international information policy, thus contributing towards standardisation, promoting sharing and reducing wasteful duplication. ICSU-AB has Working Groups concerned with the international standardisation of bibliographical descriptions, exchange formats for machine-readable text[9] and an internationally acceptable physics classification.[10] It also arranges for co-operation: among editors of primary periodicals to simplify the work of the secondary services; between the publishing services for the marketing of secondary information; etc.

Finally, ICSU was very directly involved, together with Unesco, in the launching of UNISIST[11a,b] in 1971. The original aim of creating a world scientific information system was modified in practice to the more manageable task of the co-ordination of an international network of services based on voluntary co-operation. The main emphasis at the outset was on reaching international agreement on common standards and procedures so as to facilitate exchange of both manual and machine records of information. This is, of course, a most formidable undertaking in the anarchic state of information transfer.

None the less, UNISIST has started off well; already a number of international systems for such collaboration have been successfully set up. Physics information will soon be benefiting directly, although this may not always be realised by the individual scientist.

REFERENCES

1 Weinberg, A. *Reflections on big science.* M.I.T. Press, 1967 ('The choices of big science', pp. 65–115)

2 Price, D. J. de Solla. *Little science, big science.* Columbia University Press, 1963

3 U.S. National Research Council. Physics Survey Committee. *Physics in perspective.* Vol. 1. Washington, National Academy of Sciences, 1972 (Chapter 13, 'Dissemination and use of the information of physics, pp. 890–966)

4a Anthony, L. J., East, H. and Slater, M. J. 'The growth of the literature of physics (*Reports on Progress in Physics, 32,* Pt. II, 709–67 (1969)

4b American Institute of Physics. *Physics information: a national information system . . ., 1972–1976.* New York, June 1971 (AIP/ID 71P)

5 Cooper, M. and Terry, E. *Secondary services in physics.* New York, AIP, October 1969 (ID 69-2)

6 American Institute of Physics. *A program for a National information system for physics and astronomy, 1971–1975.* New York, June 1970 (AIP/ID 70-P)

7 Coblans, H. 'An inflation of paper: a note on the growth of the literature' (*Herald of Library Science, 10,* 349–356, 1971)

8 Libbey, M.A. and Zaltman, G. *The role and distribution of written informal communication in theoretical high energy physics.* New York, AIP, August 1967 (AIP/SDD-1 rev.)

9 UNISIST/ICSU-AB. Working Group on Bibliographic Descriptions. *Reference manual for the preparation of machine-readable bibliographic descriptions,* compiled by M. D. Martin. Paris, Unesco, 1974.

10 ICSU-AB. Working Group in Physics. *World physics classification.* Draft 5 (Final) London, INSPEC, July 1972

11a UNISIST. *Study report on the feasibility of a World Science Information System,* by Unesco and ICSU. Paris, Unesco, 1971 Synopsis, 1971

11b *UNISIST Newsletter: Programme of International Co-operation in Scientific and Technical Information.* 1– , 1973– . Paris, Unesco, 1973– (quarterly, available also in a French version)

3

Science libraries

P. D. Friend

It has been something of a fashion to look with trepidation at the growing volume of scientific literature. Attempts have been made to estimate the total world output, but the variety in form of scientific and technical documents make such assessments well-nigh meaningless. One thing seems fairly certain—the total is very considerable and is growing at an ever-increasing rate. Records of the growth of physics literature of the more easily identified types—monographs, journal articles, reports—indicate an exponential growth rate, with the total doubling approximately every 8 years. Two points, however, are worth bearing in mind. First, this rate of growth has applied for a number of years: measurements of the increase in output of published pages of the American Institute of Physics show it to be exponential over the last 50 years.[1] So the problem is hardly a new one! Second, the total volume of literature, even within a comparatively narrow subject field, has now reached such proportions as to be well out of the possible range of individual possession, and so scientists and technologists are, of necessity, becoming more and more dependent on libraries.

Until comparatively recent years, the scientific libraries have been able to deal with the increase in their collections without radical changes in their handling techniques. However, the pace has quickened to such a degree that old methods, even after streamlining, are proving inadequate and entirely new systems are having to be adopted. The two developments of the last 20 years or so that have eased the problems of the librarian, are the use of computers to handle large files of document references and the ability to micro-copy the documents themselves for greater economic storage. More will be said later on how both of these techniques have been used by the librarian to his advantage. In addition to an increase in its sheer bulk, this

22

century has seen changes in the relative importance of the various forms of scientific literature. Journal articles have taken a major role in the recording of information, while research reports have increased in their importance as media for communicating scientific and technical research and development data.

SCIENTIFIC LIBRARIES

The following three questions will naturally arise and an attempt will be made to answer them: Where are the major collections of scientific literature? How do the libraries cope with the mass involved? And most important, how can the literature be consulted? Of course, only the larger national libraries have anything like full collections. In the United Kingdom the scene is dominated by the plans for the British Library,[2] and although the situation is still not entirely clear, the concept is of a reference library of scientific and technological literature, incorporating appropriate material from the present British Museum collection and that of the National Reference Library of Science and Invention, together with the massive collection that has been assembled at Boston Spa as the British Library Lending Division (BLL).[3]

Other large collections of scientific literature have centred on the universities. The Radcliffe Science Library of the University of Oxford is the science section of the Bodleian Library, with some quarter of a million scientific and technical books and nearly 6000 current periodicals, while a substantial part of the 3 million volumes and 10 000 current periodicals of the University of Cambridge Library is made up of scientific works. Other large collections in the UK having substantial scientific contents are those of the National Library of Scotland, Edinburgh, with some $2\frac{1}{2}$ million books and nearly 3500 current periodicals, and the National Library of Wales at Aberystwyth, with some 2 million books and again about 3500 periodicals. Some of the larger public libraries also have substantial collections of scientific and technical literature. Among them are the Mitchell Library of the Corporation of Glasgow, with about 150 000 volumes, and the Scientific and Technical Library and Information Service of the Manchester Central Library, with some 120 000 reference books. Further details of the holdings of, and services provided by, these libraries may be found in Volume 1 of the Aslib Directory,[4] together with references to most of the other libraries in the British Isles, including many that have specialised collections of interest to the physicist. An index to their subject specialisations is of particular value.

In the United States the pattern is similar, with the Library of

Congress in Washington dominating. It claims to hold a total of some 16 million books. Of the libraries specialising in pure and applied sciences, mention should be made of the Massachusetts Institute of Technology Library in Cambridge, Mass., with $1\frac{1}{3}$ million volumes, and the John Crerar Library in Chicago, with $1\frac{1}{4}$ million volumes, which in 1967 published its catalogue in 77 volumes.

From this it should not be inferred that the only effective libraries are the large ones. It might be so if all literature was of equal value to the user. In practice this is very far from the case, and it has been found that within any subject field there is a 'core' of publications which contains a very high proportion of the really useful literature. The distribution of 'importance' is not evenly spread but concentrated in relatively few documents. This phenomenon, first noticed by Bradford in 1948,[5] has been studied in great detail since, especially in recent years. In its simplest form the principle is well illustrated by some measurements made in a study of the distribution within scientific journals of articles relevant to the subject area covered by the *Mass Spectrometry Bulletin*—a current awareness journal produced by the Mass Spectrometry Data Centre at Aldermaston.[6] Plotting, over a period of 6 months, the number of references selected as relevant to this subject field from each journal in turn against the journals involved ranged in order of their productivity results in the curve shown in *Figure 3.1*. It will be seen that a library endeavouring

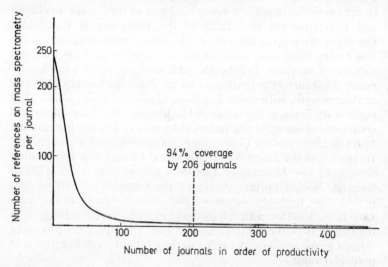

Figure 3.1

to supply the journal requirements of the mass spectroscopist would meet a substantial proportion of his needs if the first 100 journal titles listed were currently in stock. In fact, approximately 80% of the references are found in the first 100 journals and more than 90% in the first 200. In practice the librarian stocking a library to meet this subject interest will place subscriptions with as many of the journals as he is able to afford and has accommodation to store, starting the list of titles with the most prolific.

This basic principle applies to all libraries providing scientific and technical information. The scale of the curve changes, of course, with the width of the subject area, but the shape is always substantially the same and, with care, even the small library is able to stock a surprising proportion of its users' requirements. The proportion is determined by the limit of the library budget, so that titles to the left of the limiting line represent those in the library stock, while those on the right, whose references lie on the curve asymptotic with the 'journal' axis, represent the journals held in larger libraries, that are available for the occasional references they contain. This is the function performed so ably by the BLL at Boston Spa, where over 35 000 current periodical titles are held and are available for loan.[3]

HANDLING THE LITERATURE

It would be very wrong to give an impression of a modern scientific library as a mere collection of literature, however well selected. Today the role of the scientific library is to give an information service, and since this information is almost entirely in the form of printed documents, it is essential that the literature collection be as complete an anticipation of the information requirements as possible, and be housed in such a way that they can be retrieved quickly and economically. This may be something of an over-simplification, but it is a good starting point to study a little more carefully the functions involved.

CLASSIFICATION

Unless the library stock is so small that the location of any item in the store can be remembered, there must be some system of arrangement. This orderly arrangement of literature on the library shelves, generally referred to as the library classification, is in the large majority of scientific and technical libraries likely to be one of the following three long-standing systems. The library may use the Dewey Decimal system, the Universal Decimal Classification or the American Library of Congress classification system. In each case the system is an attempt to bring together, in some systematic way, subject areas that are related and so to ensure that books containing particular information are found on the shelves in reasonable proximity. Since literature can involve a multitude of relationships between all the possible subject fields, no classification can succeed in all respects, and in practice any system is in the nature of a compromise. Since all of the classifications mentioned above attempt to cater for all literature, they are all of necessity very extensive. In each case, although the over-all principle is simple enough, the width of application leads to vast ramifications, and it would be impossible to give here more than the very briefest of outline. Nevertheless, a knowledge of only the basic concepts can be of some assistance in finding one's way about a library.

Dewey decimal classification

The Dewey decimal system is more likely to be found in libraries with fairly general collections than those with predominantly scientific and technical literature. It is based on a concept of Melvil Dewey exactly 100 years ago—the first classification, incidentally, to be devised specifically for the subject arrangement of books in a library rather than as a system for grouping knowledge into subject areas. By allocating arabic numerals to the subject divisions of literature (reserving the zero for general works) he was able to break down to more and more specific sub-sections and label them by increasing the number of digits in a decimal fashion. In practice he started with a minimum of three digits for each category, dividing up the literature into 10 categories according to its predominant content as follows:

000	general works	600	technology (applied science)
100	philosophy	700	the arts
200	religion	800	literature
300	social sciences	900	general geography and history
400	languages		
500	pure sciences		

The pure sciences are then subdivided into the following headings:

510	mathematics	560	palaeontology
520	astronomy	570	anthropological and biological sciences
530	physics		
540	chemistry	580	botanical sciences
550	earth sciences	590	zoological sciences

Physics is again broken down into:

531	mechanics		photic phenomena
532	mechanics of fluids	536	heat
533	mechanics of gases	537	electricity and electronics
534	sound and related vibrations	538	magnetism
535	visible light (optics) and para-	539	modern physics

Further sub-division follows after the use of a decimal point: for example, 535 is again split into the following sub-divisions:

535.1	theories	535.5	beams and their modifica-
535.2	physical optics		tion
535.3	transmission, absorption, emission of visible light	535.6	colour
		535.8	special developments
535.4	dispersion, interference, diffraction of visible light	535.9	tables, reviews, exercise

The full schedules are in published form and will always be available in a library using this classification system.[7]

It should be remembered that the scheme is for arranging books on the shelves of a library, and so only one classification—i.e. one place on the shelves—can be allocated to a book. The librarian does his best to assess the predominant subject of the book and selects the appropriate classification number. Herein lies one of the age-old library problems. The 'predominant' subject field of a document is very much an opinion of the user, and the choice of a classification category is inevitably biased in an attempt to predict what will be the most useful aspect of the book to future library users. The compromise solution to such problems is in the use of a subject-indexed catalogue. While it is impossible to put one copy of the book in more than one place on the shelves, it is easy to include in the catalogue references in as many subject areas as might be found useful. More will be said later of the form and use that may be made of catalogues in science libraries.

Universal decimal classification

The UDC is a development of the Dewey Classification, and has been adopted by many scientific libraries both in the UK and in Europe. It is considered more appropriate than the Dewey system for the classification of technical subjects, largely because of its great flexibility and its ability to cater for very specific and precise subject divisions. The main divisions of knowledge are those adopted by Dewey, single digits being used to identify them. Thus, section 5 covers the pure sciences and section 6 the applied sciences. These are again sub-divided into units of 10, and although decimal points are not normally used before the third digit of a schedule number, it should be remembered that in the decimal sense it is implied after the first. So the major divisions of physics (53) are:

531	mechanics	535	optics, light
532	fluid mechanics, hydromechanics	536	heat, thermodynamics
		537	electricity
533	gas mechanics, aeromechanics	538	magnetism, electromagnetism
		539	physical nature of matter
534	vibrations, acoustics, sound		

Full schedules will, again, always be available in a library using the UDC system.[8, 9] However, the alphabetical indexes to these schedules are not always found to be adequate for the specific subject requirements of some libraries, and additional alphabetic guides may be provided. Maintenance of such local indexes has been eased considerably by use of the computer listing methods referred to later.

In a number of libraries the UDC has tended to become a subject catalogue classification rather than a book arrangement scheme. With the development of the long strings of digits necessary to subject classify specifically, their use as shelf indicators has led to practical difficulties. For this reason libraries have adopted modified, usually shortened, UDC schedules as their shelf classifications, and this can lead to some confusion.

One of the most powerful features of the UDC is its facility for relating any subject to any other. The appropriate subject classifications are linked by means of a colon, and any number of schedules may be related to one another in this way. Although by permutation all the possible relationships may be included in the subject catalogue, in practice the cataloguer controls this with caution.

Library of Congress

There are a few libraries in the UK that use the subject classification of the US Library of Congress, including the National Library of Wales already mentioned. By adopting letters rather than numerals, a larger number of divisions is possible at each level. Thus, the whole of knowledge is divided into 21 classes, 'science' being under section Q, which is further sub-divided into:

QA	mathematics	QK	botany
QB	astronomy	QL	zoology
QC	physics	QM	human anatomy
QD	chemistry	QP	physiology
QE	geology	QR	bacteriology
QH	natural history		

Sub-division of these sections is continued with the use of arabic numerals intermixed with letters, both upper and lower case, and decimals, and any detailed classification results in comparatively lengthy symbols. However, the system is flexible and is particularly well suited to its original purpose, that of classifying books in a library.[10]

CATALOGUE

It will be seen that all three library classifications considered above are based on a subject arrangement of the books. Public libraries generally shelve their fictional literature in alphabetical order of the authors' names. This is simply because the name of the author is the most reliable identifier of a particular novel. But public libraries generally also have a title catalogue to their collection, should a knowledge of the book's title be the only information available. In fact, the library catalogue is an attempt to list under all the headings, of either a bibliographic or a subject nature, that might be found useful in tracing a book, a description of the document sufficient to both identify it and to locate it in the library. Thus, the three essential elements of any record in a library catalogue are: (1) the heading under which the record is filed; (2) a description of the document sufficient to identify it—and to distinguish it from possible similar documents; (3) a location indicator showing exactly where in the library the document may be found.

The bibliographic headings chosen vary a great deal from one library to another. Most choose to file records under the authors or the editors concerned, as well as corporate authors or issuing organisations. There is a strong tendency in recent years to use the titles of scientific and technical literature as headings, since library users are inclined to have available only mis-spelt authors' names.

For many years the standard method of filing records of documents held in the library has been by means of 5×3 in. catalogue cards. The usual practice is for these to be of a 'unit' card nature—i.e. for each book, a sufficient number of identical cards is produced by some convenient duplication method, to enable one to be filed within each of the appropriate headings of the catalogue. Thus, the searcher is able to find full details of the required document under any of the headings used for filing, without being cross-referred to a 'main entry'. Unfortunately, 'unit' card catalogues are tedious to maintain, since any amendments must be made to all cards of the unit; and since the temptation to file under all possible headings has led to vast card catalogues, they tend to become more and more out of date.

Subject catalogues with records filed under every conceivable relevant subject category are notoriously bad in this respect. Because of the ease and speed of file maintenance by computer, the compilation of library catalogues by this means has been one of the most effective steps in recent library mechanisation. Other uses of the computer for carrying out library routine operations, such as handling orders, recording loans and listing additions to the library, have been introduced but have concerned the library user less than the new and more extensive catalogues that have evolved.

Once a library has its book records, or even a substantial proportion of them, in a form that can be computer-manipulated, it is possible to produce a multiplicity of listings under any of the headings that, by programming, can be extracted from the records. And this can be done with no further effort on the part of the cataloguer, giving him great flexibility to experiment with various forms of catalogue and to assess their value to the library users. The time and effort required to make even minor changes to 'card' catalogues has always stood in the way of the librarian's desire to adapt the catalogue to its users' needs. An example of this is to be found in the library of the Atomic Weapons Research Establishment at Aldermaston.[11] Here the records of books in the library have, for the past 6 years, been converted to computer-readable form, and this has enabled the library to print out various forms of catalogue. A form of catalogue that has proved highly popular is that produced by extracting from the book records the title of each book and listing the full details of the document under every significant word contained in its title. This is known as 'Keyword Out of Context' (KWOC) indexing; clearly an impossible task to undertake manually, but a very simple process for the computer. It has been found to be a great boon to the library user who is unable to remember the exact title of a book, but can at least be sure of one word in it! A specimen page of this catalogue is reproduced as *Figure 3.2*. It is likely that catalogues of this type will soon become commonplace. There are a good many advantages of a catalogue in computer-printed form compared with the conventional card system, apart from its easier maintenance by the librarian. It has been shown that the user is able to scan through pages of references more quickly than through trays of cards; and since the printed form is more readily reproducible, a library can have several copies of its catalogue available to the users at the same time. Naturally there are problems, the most serious being that of sheer bulk, and consequently high cost, of computer-updating and cumulative-printing the full catalogue. To be effective the latter must be constantly added to and amended, and ideally this would be done by on-line interrogation of the catalogue file in the computer,

FLOWCHARTS

Flowcharts.   Chapin N   Auerbach Publrs 1971 £5 00   179p   0877690618   681 3 06   0367a72
Library 1879|71 Library   Shelf 681 3 CHA

FLUCTUATIONS

Noise and fluctuations in electronic devices and circuits   Robinson Frank Neville Mosband   Oxford   Clarendon
Press 1974   £8 50   [7] 246p ill 24cm 0198593198   0312|75 O|O

FLUENCE

Neutron fluence measurements.   International Atomic Energy Agency IAEA 1970   184p   Oaa003295x
539 125 5 08   290|71 Library   COPY 1 Library 0387a72 SSGS Shelf 539 1 INT

FLUID

Fluid flow for chemical engineers.   Holland F A Arnold 1973   £2.20   269p   Bibliogs 0713113015
532 5*66   66*532 5   0268a74 Library 1025a73 Library   Shelf 532 HOL,

Principles of non-Newtonian fluid mechanics   Astarita   Giovanni   London [etc] McGraw-Hill 1974   £6 45
ix 289p ill 24cm 0070840229   1495a74 O|O

FLUORESCENCE

Guide to fluorescence literature. Vol.3.   Passwater R A Plenum 1974   Oaa006777x   0954a74 O|O

Practical fluorescence; theory, methods, and techniques. Guilbault G G   Marcell Decker 1973'   £16 25
664p   Bibliogs   72090964   543 426   535 372   0524a74 Library   1157a74 Library   Shelf 543 GUI.

Radiation safety for x-ray diffraction and fluorescence analysis equipment   American National Standards
Institute Subcommittee N43-1 [Washington] U S National Bureau of Standards for sale by the Supt of Docs
U S Govt Print Off 1972   £0 30   viii 11 p 26 cm "ANSI N43 2-1971 " "ANSI N43 2-1971 " 71189153   0187a74 O|O

Spectrochemical analysis by x-ray fluorescence.   Muller R O   Plenum   1972   £8 00   326p
Bibliog 0306304368 Translated from the German by K Keil 543 426 2   535 372*543 423 6   543 423 6*535
372 2041a72 O|O   1509a72 Library   Shelf 543 MUL.

FOAMS

Mechanical properties of polymeric foams.   Meinecke E A Technomic 1973   £8 50   105p   Bibliogs
0877620792   678 01*620 17   1866a73 Library 2304a72 Library   Shelf 678 MEI

Figure 3.2

both by the cataloguer for addition and amendment and by the user for consultation. At present the large-scale storage necessary is likely to be considered uneconomic for all but the very large libraries, and so compromise solutions are emerging, such as that adopted at Aldermaston. Here print-out of the various sections of the catalogue —the author index, series index, UDC index as well as the main KWOC already mentioned—are being made by COM (Computer Output in Microform). The cost of this form of computer output is considerably cheaper than line printer output, and so the catalogue can be updated and cumulated frequently. Microfiche output has been chosen as a convenient form for use in this library. Microfiches are becoming increasingly common as a medium for recording and exchanging scientific and technological data, and library users are becoming more and more familiar with them and the instruments necessary to read them. An advantage in using microfiches to record the library catalogue is the ease and low cost of making copies. Since the whole catalogue is contained in only two dozen microfiches, it is possible for many library users to have their own up-to-date copy of the library catalogue in their laboratories or offices.

A number of libraries have adopted systems similar to this, some using 16mm microfilm output, with hand-operated, or even motorised, microfilm readers for their users' access to the catalogue records. There is no doubt that the tendency will grow, and these will ultimately be the normal form of library catalogue, at least until visual display terminals, with direct on-line access to the library records in the computers, replace them.

MICROTEXT

Mention has already been made of the use of microtext to help solve the ever-present storage space problems of libraries as well as its use as a cheap and convenient form for maintaining an up-to-date catalogue. So even the briefest chapter on making the most of a scientific library would be incomplete without a few words about this form of literature.

The concept of producing documents in micro-size by reducing them photographically is a very old one, nearly as old as photography itself. However, its application as a means of reducing the volume of stored literature was not considered very seriously by librarians until after World War II. In fact, developments in library applications have been slow until comparatively recent years.[12] Microfilm, with its tedious spool-threading and earlier clumsy readers, was not readily accepted by library users. The introduction of microcards for large-scale dissemination of technical reports by US Government organisations, especially the Atomic Energy Commission, did little to endear potential users to microtext. Microcards appeared in a variety of sizes and document reductions, as photographic positives. Perhaps the most common form was the 5 × 3in. card containing up to 47 frames, each containing a page reduced some 18 to 20 times. Unfortunately, the original reports from which cards were made were themselves often of extremely poor quality, and the resulting microcards were frequently quite unreadable. In addition, the problem of illuminating such a small opaque surface to give a sufficiently bright reflected image was always a difficult one, and no readers could be regarded as really adequate. Since reports were often available only as microcards, this led to considerable frustration and a distrust of all microforms.

The development more recently of computer output in microform (COM) has led to very wide applications in the commercial world. This has naturally been followed by enormously improved microreaders of all types, and so the librarian is able to take advantage of a technique that he has seen for many years as the solution at least to some of his problems. Both microfilm and microfiche may be found in the scientific library: the film, either 35mm or 16mm, is commonly used for recording journal literature that is not consulted frequently enough to justify its storage as bound volumes. Excellent readers are now available, and most of the difficulties of handling reels of film are overcome by using ready-loaded cassettes or cartridges. In addition, reader-printers, capable of making printed copies of the screen image as required, are now available at very reasonable prices and, even more important, produce their printed sheets of a quality and cost comparable with those of photocopiers.

Microfiches are becoming even more commonly used, especially as a means of recording scientific and technical reports. They have been much bedevilled by a lack of standardisation—in over-all size, in frame reduction and spacings, and in image resolution—which has caused irritation both to the designers of readers and to librarians. However, the COSATI standard in America and the British Standard, which is based closely on it, have led to fewer variations in form.[13] Again, microfiche readers, and reader-printers, have improved and are continuing to improve at a great rate, so that many of the old bogeys of microtext have disappeared.

It should not be imagined that all the advantages of microform literature are to the librarian. Improved library services resulting from a larger available stock are clearly to the user's benefit. But there are other less obvious advantages to the user. Since microfiches are extremely easy and very cheap to duplicate, libraries tend to issue 'non-returnable' copies rather than lend them. This ensures a master copy being available in the library at all times and allows users to maintain larger personal collections of reference literature. After all, the 'information explosion' is not confined to the library.

INFORMATION SERVICES

The ultimate function of a science library is that of providing its users with the information they require, and the library's collection of literature is its main source. So, not unnaturally, one of the first steps taken by the librarian is to ensure that library users are kept informed of newly published material that is likely to be of interest to them. Known as 'selective dissemination of information' (SDI), this is done in many libraries by carefully scanning all incoming literature, whether it be books, journals or research and development reports, and bringing to the users' notice items within their subject areas. It does, of course, depend on the librarian knowing these 'customer' interests and keeping abreast of changes in them—no easy task in a large research organisation. With the present very high cost of scientific publications, the libraries' book stocks and journal title coverage, are becoming more and more limited. The library budget line of *Figure 3.1* is moving to the left. This is leading to a greater reliance by library information staff on secondary sources of references —in particular, the published abstracts journals—and since these can be very time-consuming to scan, with a number of customers' interests in mind, the present trend is to use computer search techniques to carry out this routine service.

Magnetic tapes containing full bibliographic reference data, together with various indexing systems, are available covering many subject areas. These are, with a few exceptions, the by-products of using computer methods for compiling abstracts journals or their indexes. However, they are valuable sources of references for mechanised scanning, and the information services of many libraries are now dependent on them. Briefly the technique employed is to draw up for each 'customer' a 'profile' of his interests. The form of this profile depends on the tape to be searched; but, in general, if the records include subject indexing terms taken from a specific thesaurus, it is simply a list of those terms that are relevant to his interests. On the other hand, the search can be made on the titles of the references, when the profile is a collection of words (or word roots) that are likely to appear in the titles of subject interest to the user. The process is refined by using simple Boolean 'and', 'or' and 'not' logic. Acceptance of alternative terms or words, singly or in combination with one another, as well as the rejection of titles containing irrelevant terms, is possible, and ensures an acceptably high degree of relevance in the finally selected references.

A number of these data base tapes are now available commercially, but the most important to the physicist are the INSPEC tapes, produced by the Institution of Electrical Engineers. Details of the development of these tapes and the services provided by the IEE from them have been given at some length by Barlow[13] and by Aitchison and Martin.[14] (See also p. 65.)

REFERENCES

1 Herschman, A. 'Keeping up with what's going on in physics' (*Physics Today*, *24*, No. 11, 23–39, Nov. 1971)
2 The British Library (Cmnd 4572). HMSO, 1971
3 Houghton, B. *Out of the dinosaurs: the evolution of the National Lending Library for Science and Technology*. Clive Bingley; Hamden, Linnet, 1972
4 Wilson, B. J. (ed). *Aslib directory*, Vol 1. *Information sources in science, technology and commerce*. Aslib, 1968
5 Bradford, S. C. *Documentation*. Crosby Lockwood, 1948
6 *Mass Spectrometry Bulletin*, Vol. I—1966—Mass Spectrometry Data Centre. Procurement Executive, MOD, AWRE Aldermaston, Reading
7 Dewey, Melvil. *Dewey Decimal Classification and Relative Index*. Edition 18. New York, Forest Press, 1971
8 *Universal Decimal Classification*. Abridged English Edition (3rd edition, revised 1961). BS 1000 A: 1961. British Standards Institution
9 *Universal Decimal Classification*. Second English full edition. BS 1000. British Standards Institution
   UDC 5/50, 'Exact sciences in general', 1974. UDC 53, 'Physics', 1974. UDC 54, 'Chemistry, Crystallography, Mineralogy', 1972 (51 and 52 exist in 1st edition, 1943, but are due to appear soon)
10 United States Library of Congress. Classification. (Published in classes and sub-classes in various editions. 1st–5th). Washington, Library of Congress, Card Division. 1971– . Class Q, Science. 6th edition, 1973
11 Corbett, L. and German, J. AMCOS project stage 2. A computer aided integrated system using BNB MARC literature tapes. Program, Vol 6, No. 1, January 1972, pp 1–35
12 Williams, B. J. S. *Miniaturised communications: a review of microforms*. The Library Association and the National Reprographic Centre for Documentation, 1970
13 Barlow, D. H. 'Information retrieval' (*Computer Bull.*, *16*, No. 5, 250–256, May 1972)
14 Aitchison, T. M. and Martin, M. D. 'INSPEC: Services for the physics community' (*Physics Bull.*, *23*, 523–526, September 1972)

4

Reference material and general treatises

Herbert Coblans

Workers in the pure and applied sciences have a daily need for a wide range of data and specific pieces of information. In physics its urgency increases in the overlapping fields from theory to technical physics in all its varied applications.

Thus, in a bibliographical guide it is somewhat misleading to try to include even the best of the shorter reference tools, since new and more up-to-date ones are coming on to the market and superseding all but the best of the classics. Each national tradition or language area has its favourites and it is up to the professional librarians in charge of reference departments to keep abreast of what is currently appearing. This means, of course, that in any fairly large physics laboratory the library is as important as any other section and needs more than the part-time attention of a lesser physicist.

On the other hand, it is the standard and comprehensive works of reference, the great 'Handbücher', which must be understood in their purpose, arrangement and coverage so that they can be mined for their gold. They have a special importance because their relevance is not always directly obvious. Furthermore, in this area it is not always easy to separate physics from chemistry, and all the world's great languages must be taken into consideration. Therefore it is mainly these works which will be described, and the choice will be highly selective.

The order in which they are discussed will be partly determined by form and purpose. Unfortunately, the terminology is very loose in this area and titles are often given by publishers mainly on commercial considerations. The German 'Handbuch' is particularly important

36

as a category in physics and chemistry, but it cannot always be rendered by 'handbook'. As terms 'encyclopaedia' and 'dictionary' are often confused or they overlap to a large extent in what they are made to cover.

'HANDBÜCHER'

The 'Handbuch' is a concept which aims at exhaustive and comprehensive treatment of a fairly large subject area. This means that it has come out in parts and updating is always a problem. With the rapid growth and change in physics, both theoretical and experimental, the usefulness to the working physicist of the Handbuch approach is often questioned and they are often relegated to what is casually dismissed as the 'archival literature'. On the other hand, it should be realised that this is not the place to look for the latest information, nor is exhaustivity necessarily compatible with critical evaluation, although the best of them do sometimes combine the widest bibliographical analysis with a creative synthesis, which make some of the chapters landmarks for that particular field. For example, one need only look at some of the Handbuch articles by such masters as Rayleigh, Sommerfeld, Born, Pauli and Bethe to realise their crucial impact and formative influence for the time and even later.

Handbuch der Physik (*Encyclopedia of physics*). 2. Aufl. Herausgegeben von (Edited by) S. Flügge. Springer, 1955– .
(To be completed in 54 sections, some of which contain more than one volume. In contrast to the 1st edition (1926–29) most of the articles are in English (more than 80%) or German (more than 10%), and the rest in French, with about five articles per volume.)

1, 2. Mathematical methods. 1955/56
3. Principles of classical mechanics. 3v. 1959/65
4. Principles of electrodynamics and relativity. 1962
5. Principles of quantum theory I, 1958
6. Elasticity and plasticity, 1958
6a. Mechanics of solids. 4v. 1972/74
7. Crystal physics. 2v. 1955/58
8, 9. Fluid dynamics. 3v. 1959/63
10. Structure of liquids, 1960
11. Acoustics. 2v. 1961/62
12. Thermodynamics of gases, 1958
13. Thermodynamics of liquids and solids, 1962
14, 15. Low temperature physics. 2v. 1956
16. Electrical fields and waves, 1958
17. Dielectrics, 1956

18. Magnetism. 2v. 1966/68
19, 20. Electrical conductivity. 2v. 1956/57
21, 22. Electron emission. Gas discharges. 2v. 1956
23. Electrical instruments. 1967
24. Fundamentals of optics. 1956
25, 26. Crystal optics. Light and matter. 5v. 1958/74
27, 28. Spectroscopy. 2v. 1957/64
29. Optical instruments. 1967
30. X-rays, 1957
31. Corpuscles and radiation I (in preparation)
32. Structural research. 1957
33. Optics of particles. 1956
34. Corpuscles and radiation in matter II. 1958
35–37. Atoms. Molecules. 4v. 1956/61
38. External properties of atomic nuclei. Neutrons. 2v. 1958/59
39. Structure of atomic nuclei. 1957
40–42. Nuclear reactions. Beta decay. 4v. 1957/62
43. Mesons (in preparation)
44–45. Nuclear instrumentation. 2v. 1958/59
46. Cosmic rays. 2v. 1961/67
47–49. Geophysics. 6v. 1956/72
50–54. Astrophysics. 5v. 1958/62

The Flügge Handbuch is the most important work to cover the literature of physics for the record, in all its aspects and in a definitive manner. Another publication of this kind, which has direct relevance for the working physicist, is in the neighbouring field of chemistry. While physics normally deals with matter in general, it is not uncommon that reliable information is needed on a specific element or its compounds. For this the most complete and authoritative source is 'Gmelin'.

Gmelin Institut. Frankfurt/Main. *Gmelins Handbuch der anorganischen Chemie*. 8. Aufl. Verlag Chemie, 1924–   .
It aims to cover the literature from the beginnings of scientific chemistry to 1960, and some 70 000 pages of text have appeared. It is arranged in 71 numbered sections ('Systemnummern'), mainly corresponding to the elements or in some cases groups (e.g. noble gases, rare earths), and most sections have several volumes and updating supplements. The basic volumes for all the sections are due to be published by 1975. The publishers estimate that about a quarter of the information is physics rather than pure chemistry. However, it must be stressed that some of the information can be really out-of-date in the sense that the literature coverage is governed by the year in which the volume was published. On the other hand, no significant data may well have appeared on a given compound for more than 50 years.

The above works are extensive and costly, and thus are only likely to be available in the large special and university libraries. For the student or working physicist there is the sub-category, which is strictly speaking a 'handbook'. Typical of a number of these is

American Institute of Physics. *Handbook*. Edited by Dwight E. Gray. 3rd edition. McGraw-Hill, 1972.
(It covers physics, mainly experimental, in nine condensed sections along the lines of classical sub-division. It thus represents a working tool of primary resort.)

GENERAL TREATISES

The standard treatises for each field are discussed in the chapters that follow. However, especially in physics, there is some value in the synoptic approach, summarising sometimes the outlook of a generation. They usually carry the stamp of one personality or a school of thought, and are thus integrative and logically consistent. Even though their assumptions may have been partly superseded, their treatment of the fields of physics can be very stimulating, but they are not suitable as textbooks. Selection is bound to be rather arbitrary, but one criterion is helpful: the extent to which they have been translated.

Sommerfeld, Arnold. *Lectures on theoretical physics*. Academic Press, 1949–56. 6 vols. 1. Mechanics. 2. Mechanics of deformable bodies. 3. Electrodynamics. 4. Optics. 5. Thermodynamics and statistical mechanics. 6. Partial differential equations in physics. Translation of *Vorlesungen über theoretische Physik*. Dieterich, 1947–52.
(This standard work treats physics as a unified whole, in a sense the culmination of the first quarter of the nineteenth century. In contrast the Landau and Lifshitz (below) tries to do the same from the vantage point of the contemporary period.)

Landau, L. D. and Lifshitz, E. M. *Course of theoretical physics*. Pergamon, 1959–73. 9 vols.
1. Mechanics. 2. Classical theory of fields. 3. Quantum mechanics. 4. Relativistic quantum theory (2 pts.) 5. Statistical physics. 6. Fluid mechanics. 7. Theory of elasticity. 8. Electrodynamics of continuous media.
Translation of *Teoreticheskaja fizika*. Moskva, Nauka, 1954–67.
(See p. 116 for comment.)

Feynman, R. P., Leighton, R. B. and Sands, M. *The Feynman lectures on Physics*. Addison-Wesley, 1963–65. 3 vols.

1. Mainly mechanics, radiation and heat. 2. Mainly electromagnetism and matter. 3. Quantum mechanics.
(Based on introductory lectures given at the University of California in the early 1960s, this course attempts a 'revolutionary' approach in exposition. The whole of physics, both classical and contemporary, treated from the 'modern' point of view. As Feynman says in the Preface, it is not a survey, rather it is 'serious' and in depth. Encyclopaedic in scope, it at the same time presents a picture of the physicist's universe.)

Methods of experimental physics. Edited by L. Marton. Academic Press, 1959–71. 9 volumes in 14 tomes.
(Includes volumes on atomic, electron, nuclear and molecular physics; atomic interactions; solid state, plasma physics; classical and electronic methods.)

ENCYCLOPAEDIAS AND DICTIONARIES

These commonly used reference tools are not easily categorised, mainly owing to the confusion of nomenclature. A further complication is the extent to which the subjects overlap. Thus, a general encyclopaedia will sometimes be more comprehensive and provide more information in a particular subject field than one dealing only with that subject. However, here we shall be discussing only those works which purport to cover a specific subject.

Physics is well served with a comprehensive and authoritative encyclopaedia.
Encyclopaedic dictionary of physics: general, nuclear, solid state, molecular, metal and vacuum physics, astronomy, geophysics, biophysics and related subjects. Editor-in-chief: J. Thewlis. Pergamon, 1961–   . v. 1–7 A/Z; v. 8 Subject and author indexes; v. 9 Multilingual glossaries (English, French, German, Spanish, Russian and Japanese). Supplementary volumes updating basic volumes: v. 1, 1966; v. 2, 1967; v. 3, 1969; v. 4, 1971.
(Entries, usually not longer than 2000 words, are signed and have a short bibliography. With many thousands of contributors from the UK, the Commonwealth and the USA this dictionary is reliable, kept up-to-date and thus likely to maintain itself as a standard reference work for the English language. The supplementary volumes, which contain somewhat longer articles, cover new topics and developments, and fill in the gaps revealed in the basic text. The multilingual glossary contains more than 13 000 English terms with their equivalents in the other five languages. By a system of number relationships

the English equivalents of the terms in the other languages can be found.)

International dictionary of physics and electronics. 2nd edition. Edited by W. C. Michels. van Nostrand and Macmillan (London), 1961. (This large one-volume desk tool has 15 contributing American and British editors, and contains definitions for about 15 000 terms.)

CRITICAL TABLES AND DATA COMPILATIONS

The need for evaluated, even critically selected, numerical data was already becoming urgent at the beginning of the century. It was first met by the US National Academy of Sciences with its 'International Critical Tables' (McGraw-Hill, 1926–33. 8 vols.) This pioneering work covered data up to 1931 and is thus now mainly of historical value.

Actually an individual effort to assemble such data was already made in 1883 when Landolt in collaboration with Börnstein produced the first edition in one volume of their 'Physikalisch-chemische Tabellen' (see below for details of the present sixth edition). But since World War II it has become increasingly clear that data are of such importance to science and technology that national data services are becoming necessary. The first of these was established in 1963 at the US National Bureau of Standards under the name National Standard Reference Data System. In the UK the Office for Scientific and Technical Information (OSTI) was responsible for co-ordinating activities and published 'Critical data in Britain 1966 to 1967'. Similarly, there is 'The State Service for Standard and Reference Data' (GSSSD) in the USSR. These programmes are especially relevant for national standardising activities.

While the above are responsible for the whole field, there is a large number of projects confined to narrow subjects. Particularly rich is the area of nuclear data. In fact, the whole picture is very confused, and co-ordination is essential. This is being done by the ICSU body CODATA, which has published a register of such continuing projects (see Chapter 2 above). Of the comprehensive series we shall only describe two publications, the Landolt-Börnstein and 'Tables de Constantes Sélectionnées'.

Landolt–Börnstein. *Zahlenwerte und Funktionen aus Physik, Chemie, Astronomie und Technik.* 6. Aufl. Springer, 1950– .
Band I: Atom und molekular Physik (Atomic and molecular physics). 5 vols.

Teil 1. Atome und Ionen, 1950 (Atoms and ions)
2. Molekeln I: Kerngerüst, 1951 (Molecules: nuclear structure)
3. Molekeln II: Elektronenhülle, 1951 (Molecules: outer electrons)
4. Kristalle, 1955 (Crystals)
5. Atomkerne und Elementarteilchen, 1952 (Nucleus and elementary particles)

Band II: Eigenschaften der Materie in ihren Aggregatzuständen (Properties of matter in relation to state of aggregation).

T. 1. Mechanische-thermische Zustandsgrössen, 1971 (Thermomechanical constants)
2a,b,c. Gleichgewichte: ausser Schmelzgleichgewichte, 1960–64. 3 vols. (Equilibria: excluding melting point transitions)
3. Schmelzgleichgewichte und Grenzflächenerscheinungen, 1956 (Melting point equilibria and interface properties)
4. Kalorische Zustandsgrössen, 1961 (Thermodynamic constants)
5a,b. Transportphänomene. Kinetik. Homogene Gasgleichgewichte, 1969, 1968. 2 vols. (Transport phenomena. Kinetics. Homogeneous gas equilibrium)
6 and 7. Elektrische Eigenschaften, 1959, 1960 (Electrical properties)
8. Optische Konstanten, 1962 (Optical constants)
9 and 10. Magnetische Eigenschaften, 1962, 1967 (Magnetic properties)

Band III: Astronomie und Geophysik, 1952 (Astronomy and geophysics).

Band IV: Technik. (Technology).

T. 1. Stoffwerte und Verhalten von nicht metallischen Werkstoffen, 1955 (Constants and the properties of non-metals)
2a,b,c. Stoffwerte und Verhalten von metallischen Werkstoffen, 1963–65. 3 vols. (Constants and properties of metals)
3. Elektrotechnik. Lichttechnik. Röntgentechnik, 1957. (Electrotechnology. Lighting. X-ray techniques.)
4a,b. Wärmetechnik, 1967 (Heat engineering)

Landolt-Börnstein. *Zahlenwerte und Funktionen aus Naturwissenschaften und Technik—Numerical data and functional relationships in science and technology.* Neue Serie—New series. Springer, 1961– .
General Editor: K.-H. Hellwege, Technische Hochschule, Darmstadt. (The preliminary material and, wherever suitable, the text is given in both German and English.)
Gruppe I: Nuclear physics and technology.

Band 1. Energy levels of nuclei: A = 5 to A = 257, 1961
    2. Nuclear radii, 1967
    3. Numerical tables ... angular correlations in alpha, beta and gamma spectroscopy, 1968
    4. Numerical tables for beta decay and electron capture, 1969
    5a. Q values, 1973
    5b. Excitation functions for charged-particle induced nuclear reactions, 1973
    5c. Estimation of unknown excitation functions and thick target yields for p, d, ^{3}He and alpha reactions, 1974
    6. Properties and production spectra of elementary particles, 1972
    7. Elastic and charge exchange scattering of elementary particles, 1973
    8. Photoproduction of elementary particles, 1973

Gruppe II: Atomic and molecular physics.
Band 1. Magnetic properties of free radicals, 1965
    2. Magnetic properties of co-ordination and organo-metallic transition compounds, 1966
    3. Luminescence of organic substances, 1967
    4. Molecular constants from microwave spectroscopy, 1967
    5. Molecular acoustics, 1967
    6. Molecular constants, 1974

Gruppe III: Crystal and solid state physics.
Band 1 and 2. Elastic, piezoelectric, piezo-optic and electro-optical constants of crystals, 1966, 1967
    3. Ferro and anti-ferroelectric substances, 1969
    4a,b. Magnetic and other properties of oxides and related compounds, 1970. 2 vols.
    5. Structure data of organic crystals, 1971. (a) C ... C_{13}; (b) C_{14} ... C_{120}. 2 vols.
    6. Structure data of elements and intermetallic phases, 1971
    7a–g. Crystal structure data of inorganic compounds, 1973. 7 vols.
    8. Epitaxy data of inorganic and organic crystals, 1972

Gruppe IV: Macroscopic and technical properties of matter.
Band 1. Phosphorescence of inorganic substances, 1974.

Gruppe V: Geophysics and space research (in preparation).

Gruppe VI: Astronomy, astrophysics and space research.
Band 1. Astronomy and astrophysics, 1965

In 1947 the International Union of Pure and Applied Chemistry (IUPAC) revived an earlier project to publish numerical data for chemistry, physics, biology and technology. There is no published plan for the whole series, as each volume tends to be devoted to an area of rapid growth. In addition to critical data useful values of less certainty are included as well as full bibliographies. The following are the volumes of direct interest to physicists.

Tables de constantes sélectionnées/Tables of selected constants. IUPAC, 1947– .
(The Director, P. Khodadad, in Paris is assisted by an Advisory Committee of scientists.)
v. 3. R. de Malleman. Pouvoir rotatoire magnétique (effet Faraday). F. Suhner. Effet magnéto-optique de Kerr. Hermann, 1951
v. 7. G. Foëx. Diamagneétisme et paramagnétisme. C. J. Gorter and L. J. Smits. Relaxation paramagnétique. Masson, 1957
v. 12. P. Aigrain and M. Balkanski. Semi-conductors. Pergamon, 1961
v. 16. S. Allard. Metals: thermal and mechanical data. Pergamon, 1969
v. 17. Spectroscopic data. Pergamon, 1971
(This series with its various publishers is further confused by different over-all titles—'Tables of constants and numerical data', 'International tables of selected constants' and 'Physico-chemical selected constants—New series'.)

The comprehensive data compilations are usually consulted in large libraries for rather special information where considerable accuracy or out-of-the-way material is needed. For use at the bench level each national tradition tends to have its own favourites. There are two publications which are widely used in the UK and the USA:

Handbook of chemistry and physics: a ready reference book of chemical and physical data. 54th edition, 1973/74, edited by R. C. Weast. Chemical Rubber Co., 1973.
(First published in 1914, with new editions at least every two years, this handbook has been known to generations of chemists and physicists as the 'Rubber Bible' or 'Hodgman' (its editor for many years). It is divided into five sections— A, Mathematical tables; B, Elements and inorganic compounds; C, Organic compounds; D, General chemical; E, General physical constants; F, Miscellaneous.)

Tables of physical and chemical constants and some mathematical functions. Originally compiled by G. W. C. Kaye and T. H. Laby. Now prepared under the direction of an Editorial Committee. 14th edition. Longmans, 1973.
(Although it has doubled its size—now about 400 pages—since the

previous edition, 'Kaye and Laby' provides a well-selected minimum of the most useful data needed almost daily in the laboratory. It is now entirely based on SI units.)

Although the above compilations would include definitions of most units, etc., as well as the main formulae, it is useful to be able to refer to separate collections of these.

Jerrard, H. G. and McNeill, D. B. *A dictionary of scientific units, including dimensionless numbers and scales.* 3rd edition. Chapman and Hall, 1971.
(Includes definitions of some 400 units arranged in alphabetical order, many of which provide some historical background.)

Menzel, D. H. (ed.). *Fundamental formulas of physics.* 2nd edition. Dover, 1960. 2 vols.
(31 chapters by different specialists.)

5

Bibliographies, reviews, abstracting and indexing services

R. H. de Vere, A. K. Mukherjee
and J. A. Leigh

INTRODUCTION

This chapter is concerned with the secondary literature of physics, which enables the reader to track down information in particular fields, i.e. bibliographies, review literature and abstracting and indexing services. The need for these services becomes apparent when it is realised that in 1972, 85 000 items were published in *Physics Abstracts* alone.

The secondary literature described here deals with services covering physics only or subjects peripheral to physics, e.g. chemistry, or with multidisciplinary services which include a substantial proportion of physics. Only the more important services are considered. Services on more specific topics in physics are dealt with elsewhere. Many of the services are now available in computer-readable form; these are discussed in the final section.

BIBLIOGRAPHIES

The term 'bibliography' is used to describe an ordered list of articles, reports, books, serial titles, etc.; a bibliography may be published in book or pamphlet form, as an appendix to a book, as a journal article or as a periodical publication. Review articles are a major source of bibliographies. The entries may be arranged in order of author, in subject order, in chronological order, or even in random order provided that they are indexed. A good bibliography is not a mere list; the entries should have been selected according to some criteria, and these criteria should be clearly stated at the head of the bibliography.

At one extreme we have abstract and index journals; at the other, single bibliographies, lists of citations, etc. The latter are not merely useful reading lists; they save time which would otherwise be spent in retrospective searching. The discovery of these bibliographies is facilitated by the use of guides:

Besterman, T. *A world bibliography of bibliographies.* 4th edition. Lausanne, Societas Bibliographia 1965–66. (Lists bibliographies up to 1963–64 in all subjects. Arranged alphabetically under subject headings. Includes patents.)

Besterman, T. *Besterman world bibliographies: Physical Sciences.* Totowa, N. J., Rowman and Littlefield, 1971. 2 vols. (Reprint of appropriate entries from Bestermann, T. *A world bibliography of bibliographies.*)

Besterman, T. *Besterman world bibliographies: Technology.* Totowa, N. J., Rowman and Littlefield, 1971. 2 vols. (Reprint of the appropriate entries from Besterman, T. *A world bibliography of bibliographies.*)

Bibliografiya Sovetskoĭ Bibliografii. (Vsesoyuznaya Knizhnaya Palata, Moskva. Annual; started in 1948. An annotated list of bibliographies in classified order. Two year delay.)

Bibliography Digest. Published as part of *Sci-Tech News* from 1966, Vol. 20. Chicago, Special Library Association. (Each quarterly issue of *Sci-Tech News* contains a three to four page list of bibliographies, with sections covering physics and allied topics.)

Bibliographic Index. A cumulative bibliography of bibliographies. Wilson, 1937– . Published in April and August with a bound cumulative volume in December. Lists bibliographies in English and foreign languages, in Roman script, with more than 40 entries. Bibliographies are listed whether they are separate, or parts of books or pamphlets; 1900 periodicals are also scanned for bibliographies. In the fields of pure science, agriculture and medicine, the selection is more critical than in other fields.

It is also worth remembering that the bibliographies index in a library is an important guide to immediately available material.

REVIEW LITERATURE

The essence of a review is that it bridges the gap between current literature and the treatise to come; it may serve as an introduction to a new field or as a means of keeping abreast of recent developments.

A review may cover a broad field or a narrow specialisation; it may be concerned with the latest work in a field or cover a somewhat longer period and so add historical perspective to a developing subject. Generally, the shorter the period surveyed the more is specialist knowledge required of the reader. An important feature of reviews is that they are frequently supplied with extensive, yet selective, bibliographies.

A good review should be comprehensive, critical and clear. The author should be well versed in his chosen subject. Foreign work should be well represented. Frequently it is editorial policy to commission review articles from leading authorities in the field of interest. Where and how are reviews to be found? They frequently appear in the periodical literature alongside articles reporting recent research; some periodicals are devoted entirely to review articles; many a monograph is an extensive review. Titles such as 'Advances in . . .' or 'Progress in . . .' are certain to contain some review articles. In some abstracting journals review abstracts are often specially coded. A useful guide to review literature is: *Directory of review serials in science and technology, 1970–1973*. Compiled by A. M. Woodward. Aslib, 1974. (Lists about 500 titles.)

The following list gives a selection of publications (periodicals and monograph series) devoted entirely to review articles or in which review articles are likely to be found.

Advances in Physics. London, Taylor and Francis. From 1952. Appears bi-monthly; contains one to three review articles per issue. Each issue 100–200 pages. Covers all advances in particular fields of physics.

Contemporary Physics. A review of physics and associated technologies. Taylor and Francis, 1959–   . Bi-monthly. Four reviews per issue. Cumulative index for Vols. 1–7 (in Vol. 7). This journal provides reasonably simple in-depth studies of particular topics in pure and applied physics. Very useful as background reading for professional physicists. The bibliographies are short.

Essays in Physics. Academic Press. From 1970. 1 volume p.a. Three to four articles per volume. Intended for the advanced undergraduate and professional physicist with an interest in theoretical physics. The reviews provide scholarly accounts of current thinking about the underlying concepts of physics. The bibliographies are brief.

Nuovo Cimento. Serie 1 (replaces *Nuovo Cimento.* Supplemento. Revista internazionale di fisica). Consiglio Nazionale delle Ricerche e del Comitato Nazionale dell' Energia Nucleare. Quarterly. 12

articles p.a. 10–50 pages each. In English. These reviews survey the development, over several years, of selected topics of current interest. Papers are commissioned by the editorial board and are intended for specialists.

Physics in Technology (continuation of *Reviews of Physics in Technology*). London, Institute of Physics. From 1970, 3 issues per annum. Each issue provides two or three critical reviews, each 10–30 pages in length, on particular topics in applied physics which are of current importance. Bibliographies range from useful to negligible.

Physics Reports. Review section of *Physics Letters* (section C). North-Holland. From 1970. Six issues p.a. 12–120 pages per issue. Each issue comprises one review. This review series covers recent developments in pure physics, and is intended for the professional worker in the topics covered. Good bibliographies.

Reports on Progress in Physics. London, Institute of Physics, 1934– . Appears monthly with two reviews per issue. Each issue has 150 pages. The articles are commissioned directly by the editorial board from authors of international standing. The reviews cover long-term (about a decade) developments in pure and applied physics, and are intended for professional physicists who are not experts in the subject under review. Good bibliographies.

Reviews of Modern Physics (continuation of *Physical Review supplement*). American Institute of Physics for the American Physical Society. 1929– . Cumulative indexes for Vols. 1–10, 1–27. Four issues p.a., with one to four articles per issue. These are comprehensive scholarly reviews, written for the specialist and covering all aspects of fundamental physics. Good bibliographies.

Springer Tracts in Modern Physics (continuation of and also entitled *Ergebnisse der exakten Naturwissenschaften*). Published by Springer, from 1922. Contains many authoritative, scholarly reviews of currently developing topics in fundamental physics. Intended for the specialist. Good bibliographies in English.

Occasional review articles on physics, generally of an introductory nature, may be found in multidisciplinary journals such as *American Scientist, Endeavour, Science Progress*, etc. Such articles, however, are of interest to the generalist rather than the specialist.

ABSTRACTING AND INDEXING SERVICES

Abstracting and indexing services provide ordered lists of references to help readers identify the documents they need. The entries may be restricted to mere bibliographic detail (Indexing Services) or they may, in addition, contain a description of the contents of the documents. Short, indicative abstracts enable a reader to decide whether or not he should read the original paper. Longer, informative abstracts summarise the principal arguments and data, and so to some extent act as substitutes for the original.

The bibliographic elements usually given are: (1) the title of the document, (2) the author(s), (3) the name of the publication and (4) the volume, part and page numbers and date. Additionally, (5) the title may be modified or expanded, (6) the title may be translated and the original language indicated and (7) the author's affiliation may be given.

Abstracting and indexing services may be designed for current awareness or for retrospective searching, or for both. Services designed for retrospective searching enable rapid searches to be made over a long period of time in considerable detail; this is facilitated by well-constructed classification schemes and indexes—author index, subject index, KWIC index, etc. In some cases there are separate indexes for particular forms of publication, such as reports, conferences, bibliographies, patents, etc. Indexes may appear in each issue, and there may be cumulative indexes—quarterly, bi-annual, annual; recently the major abstracting journals have published cumulative indexes covering several years. For example, INSPEC has 4 year cumulative indexes for its abstracts since 1955.

The value of a current awareness service depends on the speed with which it follows the original publications. Services restricted to current awareness do not have cumulative indexes and tend to use indicative abstracts. With the advent of computers in publishing many of the major retrospective services publish so quickly that they may also be used for current awareness. Other factors which determine the usefulness of a particular service are:

1. *Subject field.* Some services cover a wide field; others are limited to narrow topics. In some cases the services are centred on a narrow field of activity within physics, but the field of interest is wider than physics, e.g. *Nuclear Science Abstracts.*

Such services are treated in the chapters devoted to particular topics. In other cases a service may centre on a field outside physics but overlap it in certain areas, e.g. *Chemical Abstracts.* The major services of this sort are treated below. A third variation comprises those services which cover a wide field but which publish the abstracts in several different abstracting journals each devoted to a narrow topic. A good example is provided by Cambridge Scientific Abstracts

(Riverdale, Maryland), who publish several abstracting journals each covering a narrow range of topics.

2. *Form of primary publication.* The usual sources of material are journals, reports, conference papers, patents, theses, treatises, handbooks, etc. Some services, however, cover special types, such as patent specifications or conference proceedings. Some of the services dealing with particular forms (e.g. dissertations, reports, patents, etc.) are treated in other chapters.

3. *Format.* At present most services are available in printed format only, either as separate publications or as a part of other journals. However, an increasing number are available on magnetic tape (see later paragraphs). Some services are available in card format and some in microform.

4. *Size.* Quite often a service selects its entries from a limited number of sources. This may be for economic reasons or for some of the other reasons listed here.

There are many abstracting services (upwards of 1300) covering science and technology. Fortunately, several guides are available, and a careful study of these will help in identifying suitable services. Before using an abstracting journal it is advisable to read the editorial pages to ascertain those limitations which affect its potential usefulness.

Guides to abstracting services (See also references 1 and 2)

Abstracting services. Vol 1. Science, technology, medicine, agriculture. 2nd. edition (FID publication 455). International Federation for Documentation, 1969. Lists 1300 abstracting services throughout the world. Each entry gives a full bibliographic description, and an indication of size, subjects covered and the nature of the indexes, if any. An excellent guide even though it does omit several important publications.

Abstracting and indexing services centred on physics

Titles in this group embrace the whole field of physics, including theoretical and experimental physics and instrumentation, but usually excluding technical applications.

Physics Abstracts (formerly *Science Abstracts, Section A—Physics*). INSPEC, Institution of Electrical Engineers, 1898– . Fortnightly.

This is the most widely used of all the English-language abstracting journals in physics. It covers all forms of publication published in all countries, and in all languages. Currently 85 000 informative abstracts are published each year, taken from the usual sources including 2500 journals.

The entries are completely in English. The language of the original is indicated whenever an entry has been translated. Each entry consists of an identification number, title, author(s) (including affiliations), the bibliographic description of the source, the abstract itself and the number of references.

The entries are arranged in accordance with the IEE's own subject classification scheme, the details of which are exhibited in each issue; the scheme is revised from time to time.

Each issue has author and subject indexes, and there are separate indexes for bibliographies, books, reports, conference proceedings and patents. Indexes are cumulated twice yearly. Cumulated indexes for the periods 1955–59, 1960–64, 1965–68 are also available. The subject index, arranged alphabetically, contains many more terms and more specific terms than those mentioned in the subject classification. Each abstract is assigned several indexing terms besides the keywords in the titles. Some re-indexing is done by the introduction of new terms, at intervals, to minimise the separation of related papers. Names of elements, their compounds and a few special compounds are included as indexing terms.

The delay between publication of the original paper and the appearance of the abstract averages about 4 months. This makes the journal useful for both current awareness and retrospective searching.

Current Papers in Physics is a fortnightly, current awareness, indexing journal, first published in 1966 by the same body as *Physics Abstracts* and derived from the same data base. The major consequence of this is that the titles listed in each issue of *CPP* are, with some exceptions, the same as those appearing in the corresponding issue of *Physics Abstracts*; the price of *CPP* is, however, only one-tenth that of *Physics Abstracts*. The entries are arranged in the same subject order as in the parent journal. Each entry comprises title, author(s) (and affiliations) and the bibliographic details of the source publication. There are no indexes.

Current Physics Index. American Institute of Physics, 1975– . Each quarterly issue contains about 4000 abstracts of articles published in the 39 primary journals of the AIP and its member societies. Each informative abstract is accompanied by the title, authors' names and affiliations, bibliographic reference and CPM number. The abstracts are arranged by subject according to the Institute's Physics and Astronomy Classification Scheme. There is an author index in each issue; cumulative author and subject indexes are published twice yearly. (*Current Physics Index* is a replacement for *Current Physics Advance Abstracts* and *Current Physics Titles*, which were started in 1972. They covered both the AIP and a few other key periodicals in physics.)

Current Physics Titles (*CPT*) comes from the same source as *CPAA*, and like that publication it consists of three separate monthly journals; the sub-titles correspond to those of *CPAA*, e.g. *Current Physics Titles: Solid State*. It was first published in 1972.

This indexing service abstracts the 68 journals covered by SPIN (Searchable Physics Information Notices). The entries give, in addition to the usual bibliographic elements, key words and phrases, location numbers to facilitate automatic retrieval from CPM (Current Physics Microfilm), and a signal to show when a paper is a review article. The balance between theory and experiment is also indicated. The entries are arranged in the same order as in *CPAA*; full entries appear under one or more headings, and may also appear in one or both of the sister journals.

Referativnyi Zhurnal. Fizika. Moscow, VINITI, Akademiya Nauk SSSR. First published in 1954. Monthly.

The complete set of *Referativnyi Zhurnal* is very comprehensive. It abstracts 22 000 serials, 8000 monographs, 140 000 patents and standards per year, irrespective of language or country of origin, covering the whole of science and technology. All the usual forms of publication, plus manuscripts deposited but not published, are abstracted almost wholesale, and an abstract of each item appears in one or several series to which it is of interest, the bias of the abstract altering according to the scope of the series. As such, the *Fizika* series, for which source publications are not separately listed, may get items from literature in a wider field. The *Fizika* series, estimated to contain 72 000 abstracts per year, covers most of the field of physics, the detail of topics being set out in the subject classification scheme with each issue. Separate series have been devoted to some of the branches in physics (e.g. mechanics), fringe subjects (e.g. geophysics) or technical applications (e.g. electrical and electronic engineering). Mathematical physics has been placed in *Matematika*. When in doubt about the distribution of topics between related series, an ancillary publication, *Rubrikator Referativnykh Izdanii SSSR* (VINITI, Moscow) may help. This is a subject classification covering the whole of science and technology. The topic in question should be identified in this scheme, which will direct the reader to the appropriate series of *Referativnyĭ Zhurnal*.

Each abstract includes: (1) The title of the article in Russian as well as in the original language; the language of the original is indicated. When the original language is not written in Roman characters, it is transliterated into Cyrillic script. (2) All collaborating authors in Cyrillic or Roman characters as they appear in the original. (3) The abbreviated title of the publication. (4) The name of the abstractor. The abstracts are in Russian and are fairly informative, sometimes giving important equations and results. The entries are grouped, according to VINITI's subject classification scheme, into seven sections, each of which is sub-divided into two further levels. Each section is separately available.

There is an author index at the end of each section of each issue in Cyrillic and Roman sequences; there is no combined index for the issue. There is a single annual author index also in Cyrillic and Roman sequences. All collaborating authors are mentioned.

There is no alphabetical subject index in each issue. Abstracts on a topic may be found from the table of contents, which mentions the abstract number where the topic starts. In the annual subject index, headings and sub-headings are alphabetically arranged. An abstract is referred under one or more headings by analysis of its contents.

The interval between the publication of the original and the appearance of its abstract ranges from 3 months to 1 year, no significant difference being noticeable between Russian and foreign articles. The annual index follows about a year after the completion of the volume. Though *Referativnyĭ Zhurnal* is not recommended for current awareness, it is perhaps the most up-to-date source of information on the state of science and technology in the Soviet bloc, and the only service to signal many publications which are not covered by other abstract journals. It is useful for retrospective searching because of its well-organised classification, detailed indexing and wide coverage. A brief introduction to the use of this journal is given in *A guide to Referativnyĭ Zhurnal*, by E. J. Copley.[3]

USSR and East Europe Scientific Abstracts. Physics and Mathematics (prior to 1973 appeared as two publications—*USSR Scientific Abstracts. Physics and Mathematics*). Arlington, Virginia, Joint Publications Research Service. Irregular, about one issue per month.

This English-language abstracting service publishes a few selected articles from the USSR and Eastern Europe. Besides abstracting from the usual sources, it includes translated items from *Referativnyĭ Zhurnal*—full bibliographic references to these Russian abstracts are given. The coverage of mathematics is about 25%, but varies from zero to 50%. It seems to cover most branches of theoretical and experimental physics, but the topics covered change from issue to issue. At present each issue has two sections—one for the USSR and one for Eastern Europe. In the East European section there is a further sub-division by country of origin—East Germany, Poland, Romania, Hungary, Bulgaria and Yugoslavia have been noted.

The abstracts are informative, and in some cases extensive—two or three pages in length and containing equations, diagrams and tables. Each entry quotes all collaborating authors (and affiliations), title of paper, place of publication, transliterated title of source publication (abbreviated in most cases and sometimes translated into English as well) and, in a few cases, the name of the abstractor. The entries are classified under a few simple headings. The time lapse between publication of the original and the appearance of its abstract varies from a few months to over a year. There are no indexes, which makes retrospective searching difficult. The usefulness of the abstract lies in signalling in English the existence of important Soviet papers to readers who would not use *Referativnyĭ Zhurnal*. The number of abstracts, estimated at 80 per issue, is a small fraction of the literature in the field.

Physics Express. New York, International Physical Index. First published in 1958. 12 issues per year.

This is an English-language digest of current Soviet literature in physics and fringe subjects such as geophysics, biophysics, astronomy, etc. Sources are periodicals, mostly Russian, and conference papers published in periodicals. Seventy-two periodicals seem to be fully abstracted, and 23 in related fields are scanned for selected items. The title of the article and its abstract are in English only. The first author is mentioned in most cases. The length and nature of the entries vary—titles only, copies of author's introduction or summary in English, translations of author's summary or excerpt, informative abstracts running into several pages with diagrams, equations and deductions, even complete translations appear in earlier issues. The number of entries is estimated to be 12 000 per year. The abstracts lag behind the original by a year or so.

The entries are grouped under 27 major sections. There is an author index in each issue, but no annual index. A cumulative index for the years covering 1958–63 is available. In its subject index, entries are arranged under 19 major sections and then chronologically according to volume and issue number within each section. Retrospective searching is difficult because of this insufficient indexing. The digest renders useful service by making Russian literature available in English. As the coverage is exhaustive and sources are listed, the searcher knows what to expect; it is not a hit-or-miss situation as in *USSR and East Europe Scientific Abstracts.*

Kagaku Gijutsu Bunken Sokuhō: Butsuri ōyō Butsuri-hen. English title: Current Bibliography on Science and Technology: (Physics and Applied Physics). Tokyo, Nihon Kagaku Gijutsu Jōhō Senta (The Japan Information Center of Science and Technology). First published in 1959. Semi-monthly.

This comprehensive Japanese service covers the whole of science and technology. The physics series contains about 50 000 abstracts per year of articles selected from over 200 Japanese, 100 Cyrillic and 800 Roman-letter journals, 30 different report series and 50 conference proceedings published in all major languages. It covers all branches of physics, but little attention is paid to technical applications or applied mathematics.

Each entry contains the title of the article in Japanese, a code for the country of publication, a code for the language of the original, the title of the article in its original language (in Roman or Cyrillic characters), all collaborating authors in Roman, Cyrillic or Japanese characters as they appear in the original; the number of illustrations and references, and the usual bibliographical elements; also mentioned are: the UDC class-mark(s), a code indicating the type of article (e.g. original, review, etc.), and a code indicating whether the same abstract appears in other series of this abstracting journal. All numerals are Arabic. The abstract is in Japanese; it is indicative, consisting of a few lines. Each entry is located in its appropriate class according to a subject classification scheme, consisting of 22 headings divided into 51 sub-headings. A multifaceted entry is described in full at one place and cross-referred from other headings.

In each issue there is a subject index only, consisting of keywords from titles listed according to the 'goju-on' system. At the end of each volume there are subject and author indexes, and three lists of source publications—for journals, conferences and reports. The subject index is not alphabetical but is a subject classification scheme devised by JICST. The author index mentions all collaborating authors in three sequences: in Roman, Cyrillic and Japanese scripts as they appear in the abstracts. The abstracts are delayed from several months to over a year. Although it is not used for current awareness (for which see *Zasshi Kiji Sakuin, Kagaku Gijutsuhen*), it is the most important source of information on the state of physics in Japan. Good indexing makes it useful for retrospective searching.

Bulletin Signalétique. Physique. Centre National de la Recherche Scientifique. 1956– . Monthly; occasionally two issues are combined into one.

The complete set of *Bulletin Signalétique* contains 500 000 references per year from 9000 periodicals published in all major languages all over the world, embracing the whole of science and technology. The Physique series contains 19 000 entries per year selected from the usual sources. Other series related to physics are: Astronomie; Physique spatial; Geophysique; Electronique; Electrotechnique; Physique, chimie et technologie nucleaires; Cristallographie; Physique de l'état condensé; Physique atomique et moleculaire. In a separate issue (1973), called *Système P.A.S.C.A.L. Accès à l'information,* the classification of topics in all the series is shown; an alphabetical index helps to locate any particular topic in a series.

Each entry contains: (1) The names of all collaborating authors and their affiliations. (2) The title of the article in the original language in the case of English, French and German; titles in other languages are translated into French—the original language is indicated. (3) The name of the country. (4) The number of references, if any. The abstract is in French and is indicative in most cases. Entries are grouped according to a four-level hierarchical classification. Cross-references to abstracts placed under other sections are provided.

In each issue there is a table of contents giving the page numbers of topics according to the classification scheme, an alphabetical subject index and an author index incorporating all collaborating authors. The indexing terms are obtained from an analysis of the contents of the papers, and not just from the titles. From 1973 the subject index for mechanics is separated from the rest in each issue. Subject and author indexes are cumulated annually within a year after completion of the volume. The interval between the appearance of a primary publication and the appearance of its abstract varies from a few months to over a year; abstracts of some French articles appear rapidly. This abstract journal is suitable for retrospective searching, especially for French literature.

Physikalische Berichte. (Formerly *Die Fortschritte der Physik.*) Vieweg, 1845– . Edited by Deutsche Physikalische Gesellschaft E.V. and Deutsche Akademie der Wissenschaften zu Berlin.

This monthly abstract journal is the major source of German literature in physics. It covers all branches of physics and such fringe areas as biophysics, geophysics, etc. Material is abstracted from books, journals (over 900 titles are scanned), reports and conference papers, from all countries and in all languages; at present patents are excluded. Currently (1973) about 55 000 abstracts appear each year. Each entry consists of a title (in the original language; it is transliterated into Roman script if necessary), author (and affiliation) and the bibliographic details of the source. The abstracts are in German, informative and signed. The entries are arranged in subject-classified order according to a scheme devised for

Physikalische Berichte; there are 12 main subject headings which are further sub-divided. Details of the classification scheme appear in each issue.

There is an author index (but no subject index) in each issue; a cumulated author index appears annually. There is an annual subject index arranged primarily in subject class order. Additionally there is a keyword index of subject headings, published in the annual subject index and also in issues 1 and 7 each year. The indexes and relatively rapid abstracting make this journal useful for both current awareness and retrospective searching.

Resúmenes de artículos científicos y técnicos. Serie B, Física aplicada. Madrid, Centro de Informacion y Documentacion. First published in 1964. Monthly; occasionally two issues are published together.

This is the only one we have come across in the field of physics with an emphasis on technical applications. In the most recent distribution of subjects, Series B, in its two sections, comprises energy technology (nuclear, thermal and hydraulic), electrotechnology, measuring techniques, electromagnetic waves, electronics, telecommunications, cybernetics, and optical and acoustical recording. The abstracts are indicative. The titles of the articles and the abstracts are in Spanish; the language of the original is indicated by a code. Indexing terms and the organisation responsible for abstracting the article are given with each entry. Series B is estimated to contain 10 000 abstracts per year from over 350 domestic and foreign journals in the major languages. The journals chosen reflect their technical bias in many cases, but many titles show that they are of interest to pure science as well. There are monthly and annual subject indexes with alphabetically arranged indexing terms for the two sections of Series B combined. There is no author index.

Major services covering science and technology

A physicist, though specialising, needs to keep in touch with a wider field. Articles relevant to his work may appear in publications not normally scanned by physics abstracting services. Physics literature alone may not cover borderline subjects such as biophysics, geophysics, crystallography, meteorology or medical physics. All the sciences enrich by mutual interaction; the methods, techniques and instrumentation of physics find applications in all sciences, and many fields of study are interdisciplinary, e.g. cybernetics, information theory, oceanography, space research, etc.

Science Citation Index (SCI). Philadelphia, Institute for Scientific

Information. First published in 1961. There are three quarterly issues, the fourth is incorporated in an annual cumulation; 5 year (1965–69) cumulations of *Citation Index* and *Source Index* are available.

This is an international interdisciplinary index to about 2500 journals, embracing the whole of science and technology relating references to sources. By reference we mean an item cited in an article; by source the article which cites it. *SCI* treats each journal comprehensively by picking out every reference from text, footnote or bibliography of all source items in the form of regular articles, editorials, critical reviews, letters, meeting reports, etc., and arranging them into three indexes. The reference year may be any year in recorded history; the source year is a very recent one.

Citation Index (*CI*) lists all items cited during the current year alphabetically by cited first author with year, journal, volume and starting page. Under each reference citation all source citations are listed, giving citing first author, journal, type of source item, volume, starting page and year. Cited reference publications may be journals, books, patents, reports, conference proceedings, etc. Anonymous Reference Citations follow the letter Z in *CI* in order of journal abbreviations. This is followed by Patent Citation Index in order of patent number.

Source Index (*SI*) is the complete author index to the current literature arranged alphabetically by all citing authors with full description of each citing item under the first author, all secondary authors being cross-referred to the first. Each entry contains these elements in order: first author, co-authors, code for language other than English, title of article (translated into English when in other languages), journal title, volume, page, year, type of source item, number of references contained and issue, part or supplement number of source journal. Anonymous items are at the beginning of *SI* arranged alphabetically by journal abbreviations. *SI* includes a Corporate Index arranged alphabetically by organisations to which the authors are affiliated.

Permuterm Subject Index (*PSI*) is a permuted keyword index to all articles processed for *SCI*. Every significant word (primary term) in a title is paired in turn with every other significant word (co-term) to produce for each annual index more than 10 million word-pairs which are alphabetically arranged, giving under each pair the first author which provides an entry into *SI* for full details.

Science Citation Index is a unique bibliographical tool for the scientist and librarian. Although the three indexes are most effective when used together as a total retrieval system, the combination of *CI* and *SI* can be used for citation and author searches or biblio-

graphic verification, and the combination of *SI* and *PSI* for subject
and author searches. Starting with a reference belonging to an earlier
date, source articles of a later date may be found through *CI* and *SI*,
thus moving forward in time. Using these source bibliographies,
earlier references may be traced, moving backward in time. When no
starting reference is known, *PSI* (current or old) will provide an entry
into source articles on a particular topic. Successive application of
this process, called cycling, will assemble a large collection of related
papers on a subject, although there will be some irrelevancy or
'noise'.

Citation indexing is a novel method of indexing. A citation implies
a relationship between a part or the whole of a cited paper and a part
or the whole of the citing paper. The subject of a search is symbolised
by a starting reference rather than by a word or subject heading. No
artificial language is needed to describe the content of a document,
but the indexing is based on the association of ideas, thus spanning
the gap between the subject approach of the indexer and that of the
searcher.

The basic question *SCI* answers is: where and by whom has this
paper been cited? But the variety of applications to which it can be put
is limited only by the user's imagination. *SI* may yield other relevant
current articles by a known author. Scientists and institutions currently active on special problems may be identified. It facilitates
feedback in the communication cycle. Scientists may want to know
whether their work has been applied or criticised by others. Because
corrections are indicated, it is an aid in following particluar articles.
With several years' *SCI* at hand, historical research may be continued
over a period: to determine whether an idea is really original; where
and how it has been applied; whether it is sustained, confirmed,
rejected, modified or absorbed into later work. A review on a topic
with a bibliography, however complete when compiled, becomes
outdated with time. Citation indexing permits these references to be
followed up, updated and reorganised. An evaluation of scientific
papers can be made from the number of citations to it: the importance
of emergent ideas in science and their influence on subsequent research can be estimated. An evaluation of journals can be made, for
selection for a library.

SCI Annual Guide and Journal Lists (1972) gives much useful
information; it shows clearly how to use the indexes and describes
many possible applications. With its emphasis on multidisciplinary
journals and its broad coverage *SCI* signals articles on a topic from
journals not usually devoted to it, and so helps in interdisciplinary
study. With minor exceptions each quarterly or annual issue tries
to include all issues of journals published and available during the

period covered. It is suitable for retrospective searching and to some extent current awareness.

Current Contents. Physical and Chemical Sciences. (Earlier sub-titles: *Physical Sciences*, and separately *Chemical Sciences*.) First published in 1961. Weekly.

Because *SCI* comes out at long intervals, the same organisation provides this service specially aimed at current awareness. This journal consists of a reproduction of the tables of contents of selected journals, from all countries; these contents lists are arranged under the headings: Physics, Nuclear Science, Earth Sciences, Crystallography, Metallurgy, Mathematics, Instruments, Multidisciplinary, Chemical, Space Science. For difficult languages, e.g. Slavonic, the contents are translated into English; the title of the journal is in both Cyrillic and Roman script. For other European languages the original is given, unless the journal provides titles in English. The language of the articles and abstracts, if any, is mentioned. The declared policy is to publish the contents pages 3 to 4 weeks after they are received; extracts from the previous month's issues seem to be common, although extracts from issues published 8 months previously as well as from the current month have been noted. Advance extracts from issues to be published have also been observed.

Not all types of physics journals are covered; the emphasis seems to be on those covering the whole or most of physics, but there are few on narrow topics. Purely technical journals are not included unless they have sufficient material of interest to pure physics. There is a sub-series: *Current Contents: Engineering and Technology.* Emphasis is on journals publishing original work and research results.

Each issue lists over 150 journal titles arranged alphabetically. There is an author index and address directory (first authors only). In the Triannual Cumulative Journal Index there is a complete list of about 750 journals covered by the service. Since May 1973 a subject index has appeared in the form of an ancillary journal, *Weekly Subject Index*, which picks out keywords from the titles.

British Technology Index. 1962– . London. The Library Association. Monthly, cumulated annually.

This is an index of major articles selected from over 300 British technical journals. Although the emphasis is on technical aspects, many articles are of interest to pure science as well. An outline of the subject fields covered mentions electronics, applied optics, applied acoustics, heat, engineering thermodynamics and mechanics, along with other technical topics. Instruments of all kinds are included.

Indexing terms chosen from an analysis of the contents of a paper are arranged in a string with one main heading and several modifiers;

full bibliographical details of the paper appear at one place; permuted strings where each of the modifiers in turn becomes the main heading are all cross-referred to the principal string. The specificity of the index string generally makes it co-extensive with the subjects of the article; as such, few articles are listed under a string, but each article can be retrieved from various approaches. Besides such permutations, general cross-references are abundant. The number of distinct titles is estimated to be 24 000 per year. There is an author index in each issue. The annual cumulation follows 6 months after the completion of the volume. The previous month's articles are frequently indexed, the maximum delay being not more than 4 months. It can be used for current awareness as well as retrospective searching.

Applied Science and Technology Index (formerly *Industrial Arts Index*). 1913– . Wilson Company. Monthly (except July); every 3rd issue cumulates the previous two, and there is an annual cumulation.

This indexes apparently all the papers in about 250 journals, 95% of which are American, the rest British or Canadian. Their titles are mostly appropriate to the scope of the field covered by applied science and technology, but the articles on physics, both theoretical and experimental, are often indexed. The average delay in indexing after publication of the original is about 6 months. The annual cumulation follows in about 10 months.

Services peripheral to physics

Here we discuss, briefly, the three major abstracting services closely associated with physics, likely to interest the physicist.

Mathematical Reviews. 1940– . Providence, Rhode Island, The American Mathematical Society. Monthly.

This abstracting journal, besides covering all topics on pure mathematics which a physicist may need as a tool, also covers applications of mathematics in various branches of science. It contains over 16 000 items per year. There is an author index, a key index and a Subject Classification. The key index is a short list of special publications such as reports, conference proceedings, etc., and not a keyword index. There is no subject index. The subject classification scheme lists in detail many narrow topics under major subject headings, but individual abstracts are not mentioned; monthly issues have to be searched under the major heading to find abstracts on a narrow topic. Cumulative author indexes, each covering several years, are available.

Chemical Abstracts. 1907– . Columbus, Ohio, American Chemical Society, Chemical Abstracts Service.

It is needless to emphasise here the intimate relationship between physics and chemistry. There are many overlapping fields, some of which may be better covered in the chemical literature. *Chemical Abstracts* is the most comprehensive service available, abstracting all fields of chemistry and chemical technology and also including related topics; it publishes over 300 000 abstracts per year. Currently there are 80 sections grouped under major fields as follows: Biochemistry (1–20), Organic Chemistry (21–34), Macromolecular Chemistry (35–46), Applied Chemistry and Chemical Engineering (47–64), Physical and Analytical Chemistry (65–80). Each group is available separately and carries a keyword index to the issue; since 1967 it has been appearing weekly, with sections 1–34 and sections 35–80 appearing on alternate Mondays. Within each section the abstracts are arranged by form of publication—published papers, books and monographs, patents; there are cross-references to papers in other sections.

Each issue contains an author index, a numerical patents index, a patent concordance and a keyword Index (from title and text of abstract). Half-yearly cumulative indexes consist of author index, numerical patent index, patent concordance, subject index (in two parts, general and chemical substance index), a formula index, an index of ring systems, and a Hetero-Atom-In-Context (HAIC) index. Decennial indexes covering the period 1907–56 and quinquennial indexes covering the period 1957–71 are available, the latest being the eighth collective index (1967–71). The collective formula index (1920–46) and the collective patent number index (1907–36) have also been compiled. The *Introduction to Index Guide* is recommended for the efficient use of these abstracts. Because of the completeness of the indexing, *Chemical Abstracts* is very suitable for retrospective searching; it is not a current news service, for which there is a semi-monthly service, *Chemical Titles*.

Biological Abstracts 1926– . Philadelphia, Biosciences Information Service. Semi-monthly.

This series, containing over 140 000 abstracts a year, is the most important in the life sciences. Basic medical sciences are well covered but not the particlarly clinical aspects. Fairly informative abstracts are arranged according to a scheme with broad headings and sub-headings, of which there are more than 500 in alphabetical order at both levels.

At present each issue has four indexes: author, BASIC, systematic and CROSS, each cumulated annually shortly after each volume is completed. BASIC (Biological Abstracts Subjects In Context) is a KWIC index.

Since Vol. 40 additional keywords have been added, and now an average of five terms are added to each title to overcome the inadequacy of the KWIC index. CROSS (Computer Rearrangement of Subject Specialities) lists the cross-references to abstract numbers placed under other sections. In the biosystematic index abstracts are listed under each taxonomic category. The monthly *Bio Research Index*, an ancillary journal, is a KWIC index to peripheral material not abstracted in *Biological Abstracts*.

Computer-based services

Most of the major abstracting and indexing journals are now available in computer-readable (magnetic tape) form. The tapes were originally produced as by-products of the printing process but are now recognised as information media of major importance. Several works describe the tape services available.[3-6]

The magnetic tape forms of abstracting and indexing journals have many advantages. They are often available in advance of the printed version. They can be searched rapidly and in a variety of ways. In particular, more thorough subject searching is possible either by means of additional subject terms not present in the printed version or by the free text searching of titles or abstracts. However, fairly extensive computing resources are required to make use of them, which limits their use to information centres and large organisations, who in turn provide services to the individual. Some centres—such as the Royal Institute of Technology, Stockholm; the University of Georgia; or the National Science Library, Canada—provide services from a number of different tapes. This is very satisfactory for the inquirer whose search requirements can be matched against whichever of the data bases are most appropriate to his subject needs. Another trend is that towards on-line services. Lockheed Information Systems, the Search Service of the Systems Development Corporation, and, in Europe, the Recon system of the European Space Agency are services of interest to the physicist. All three cover a wide range of scientific data bases.

Services most commonly provided by information centres using magnetic tapes are for current awareness; they may be tailor-made SDI services to suit the requirements of an individual or standard profile services aimed at a group of workers in specific fields. Such services may also be offered by the publishers of the tapes, who generally produce a package of interrelated services.

INSPEC, the publishers of *Physics Abstracts*, provide biblio-graphic information in printed, microform and magnetic tape forms, and a range of current awareness services. Their services are derived from INSPEC's computer-readable Current File. There are two forms of the magnetic tapes. INSPEC 1 contains the full details, including the abstracts, and is available in sections corresponding to the three abstracting journals. It is possible to subscribe to just the physics section. INSPEC 2, which is available earlier, does not con-tain the abstracts, but does include the full bibliographic details and free index terms, so that thorough subject searching is possible. INSPEC 2 is not divided into sub-sections and also covers the fields of computers and control, and electrical and electronic engineering. The physics section of INSPEC 1 is roughly the same basic price as INSPEC 2. Backfiles are available to 1969 and sample tapes can be obtained. INSPEC also offer two current awareness services—an SDI service and a standard profile service called Topics.

The other main purveyor of bibliographic information in physics is the American Institute of Physics, whose tape service is called SPIN (Searchable Physics Information Notices). SPIN is part of a group of AIP services in the Current Physics Information program, which also includes *Current Physics Index* and *Current Physics Micro-form* (CPM). SPIN, limited to covering some 40 core physics journals, contains about 20 000 items annually. SPIN records contain very full details, including full bibliographic information, the abstract, the citations and the CPM numbers.

Because of their broad coverage the tapes of *Nuclear Science Abstracts* and the International Nuclear Information Service (INIS) have general relevance. *NSA* tapes are available by purchase with no restrictions from the USAEC. INIS tapes are obtainable free but only by international organisations and the governments of member countries participating in INIS. In both services the INIS thesaurus is used as a basis for the addition of subject descriptors. Neither tape includes abstracts; INIS provides these on microfiche. *NSA* has of course been going longer than INIS and backfiles are available to 1962. The publishers of *NSA* provide a retrospective search service. The following services, although not primarily intended to deal with physics, cover a substantial proportion of the physics output and are also important for interdisciplinary topics.

Chemical Abstracts Condensates has been compared favourably with SPIN in its coverage of physics.

Chemical Titles is another CAS service of interest. It appears 2 months ahead of CAC but it has a much more restricted coverage and subject searching is possible only on titles, whereas CAC includes free index terms.

The magnetic tape of the multidisciplinary *Bulletin Signalétique* has been available since 1972, through PASCAL. The monthly tapes are issued in 50 sections corresponding to the subject divisions of the journal. Each entry includes descriptors from a controlled thesaurus and a short abstract in French. An SDI service is available from the publishers, CNRS.

The Institute for Scientific Information, the publishers of *Science Citation Index*, supply two tapes of interest to the physicist. The *ISI Source Index* tape contains source articles. There are about 8 000 entries per week. No indexing terms or abstracts are present, but the tapes are available very soon after the publication of the source literature. Both source and citation data are included in the *ISI Source and Citation* tapes. In 1974 the citation tape included 5 million cited references from 400 000 sources. For the individual the ISI have an SDI service called ASCA (Automatic Subject Citation Alert), in which profiles can be constructed from cited publications as well as from words or phrases in the titles of articles; and a standard profile service, ASCATOPICS.

Another interdisciplinary service is *Pandex Current Index to Scientific and Technical Literature*, published by the CCM Information Corporation, which is available in printed, microfiche and magnetic tape form, and includes about 240 000 items annually from 2400 journals. Keywords from a controlled thesaurus are added for subject searching. SDI programs are available for the tape supplier.

Government Reports Announcements of the NTIS is available as a twice-monthly magnetic tape file. Over 150 Governmental agencies (including USAEC and NASA) provide data for the file. Records contain abstracts and are indexed by descriptors from the *Thesuarus for Scientific and Engineering Terms* (but descriptors assigned by other agencies are accepted). Programs for SDI and retrospective searching are available. The NTIS also run a retrospective search service called NTISearch and a standard profile service for the supply of documents SCIM (Selected Categories in Microfiche).

Other services in which the main emphasis is on a different discipline but which may be relevant to the physicist are the computerised forms of *Engineering Index* (called *Compendex*) and *Biological Abstracts* (*BA Previews*). Note that the latter does not contain abstracts.

Finally, a service of a different kind which may supply information in advance of any secondary literature source. The Smithsonian Science Information Exchange maintains a fully indexed computer-readable file of research projects supported by the US Government.

Searches are carried out on demand and indicate the research work planned or in progress in a particular field, perhaps years before any results are published.

REFERENCES

1 *Ulrich's International periodicals directory.* Bowker, 1972
2 Mukherjee, A. (ed.). *Abstracting and bibliographical journals held by NRLSI.* London, National Reference Library of Science and Invention, 1972
3 Copley, E. J. *A guide to Refertavnyi Zhurnal.* 2nd edition, revised. London, National Reference Library of Science and Invention, 1972
4 Finer, R. (comp.). *A guide to selected computer-based information services.* London, Aslib, 1972
5 Herner, S. and Vellucci, M. J. (ed.) *Selected federal computer based information services.* Washington, Information Resources Press, 1972
6 Schneider, J. H. *et al.* (ed.). *Survey of commercially available computer readable bibliographic data bases.* Washington, ASIS, 1973

6

Special fields: patents and translations

Felix Liebesny

PATENTS

Legal aspects

The whole area of patents is one that is far too often shrouded in mystery for many people, even those to whom it should be of considerable interest and importance. The reason for the apathy or even antipathy which people feel for patents is most likely to be found in the rather obscure and legalistic language in which these documents are couched and also the rather drab and dull form in which the patent literature is so often presented.

Patents are of considerable antiquity and can certainly be traced back as far as about 500 B.C. to the luxury-loving Greek colony of Sybaris, where, according to the Greek historian Phylarchus, 'if any confectioner or cook invented any peculiar and exclusive dish, no other artist was allowed to make this for a year; but he alone who invented it was entitled to all the profit to be derived from the manufacture of it for that time, in order that others might be induced to labour at excelling in such pursuits'.

This inspiring quotation highlights the basic principles of the patent system, namely that a patent must relate to an invention (not a discovery) and that a monopoly is granted thereon which has certain limitations such as a time limit, which in that particular case was 1 year only. The patent monopoly is one of the few permitted under the current curbs of legislation such as the British Monopolies Act of 1956 or the Sherman Anti-Trust legislation of the United States of America. The reason for the exception may lie in the fact that the patent monopoly represents a barter deal in which the state grants some exclusive rights to the inventor for which he in return discloses his full knowledge of the subject of the invention.

Many of these points are enshrined in the cornerstone of many patent systems, namely the Statute of Monopolies, proclaimed by Elizabeth I in 1601 'for the reformation of many abuses and misdemeanours committed by patentees of certain privileges and licences'. This led to the establishment of the oldest still-extant patent system, that of the United Kingdom, and its Patent Act of 1623.

In that Act an invention was defined as relating to a novel manner of manufacture; this concept has since been expanded by learned judges to also encompass products resulting from such manners of manufacture and methods for testing the manner of manufacture or the resulting products. What constitutes a novel manner of manufacture has been the subject of many cases of litigation, and an interesting recent dispute centred around the question whether computer software was a manner of manufacture or not.

The scope of the definition is, however, still very wide and therefore many inventions are made every year. In the UK more than 60 000 applications are filed annually for patents which eventually result in the publication of more than 40 000 printed patent specifications. Nobody would claim that all are really world-shaking contributions to the advancement of science and technology or indeed to our daily life. Yet the number of patents for which renewal fees are paid for the whole of their life—perhaps not a very reliable yardstick for their usefulness—is still about 1 in 10. Another interesting measure of useful contributions by patents is a collection compiled by the staff of the Science Reference Library of the British Library (previously the National Reference Library for Science and Technology, and before that the Patent Office Library) of some 50 patents published during the last 200 years which are deemed to have made a substantial contribution to technical progress.

Thus, in the view of these selectors there were only 50 really noteworthy contributions during that period and some of these can now be considered to be rather obsolete. This collection contains the patents by James Watt (steam engine), Auer von Welsbach (gas mantle), Hollerith (punched card), Baekelund (Bakelite), Logie Baird (television) and Sir Frank Whittle (jet engine).

But these inventions were quite outstanding developments and cannot really be placed on the same plane as the numerous technical improvements which, while not outstanding, yet make significant contributions to the advancement of technology.

Under current British practice, which is governed by the Patents Acts of 1949 and 1961, an invention may be protected by filing a complete specification which may be preceded by a provisional specification filed up to 12 months previously. The filed specification must contain a full description of the invention, preferably accompanied by drawings and examples; the specification must end up with one or more claims which define the scope of the protection being sought. The examiners in the Patent Office will submit the application to careful scrutiny with respect to patentability and novelty; this concept of novelty as interpreted by the examiners relates to disclosures mainly in the British patent literature during the preceding 50 years. If the examiner cannot find any objections on those points the application will be accepted, given a final seven-figure number and laid open to public inspection for a period of 3 months during which any interested member of the public may lodge opposition to the grant of a patent on the application.

The grounds for lodging opposition include the following:

1. Lack of novelty, which may be based either on prior publication or prior public use; 'prior' in this sense means prior to the earliest date which appears on the specification, which may be the date of filing the provisional or complete specification or of a foreign application. Any evidence to be brought must of course be fully substantiated.

2. Obviousness, by which is meant that the inventive step described in the specification is obvious to a rather mythical being, the 'man skilled in the art', who is clearly a relative of the equally mythical 'man in the street' so beloved by politicians and newspaper leader writers.

3. Insufficiency, where again the 'man skilled in the art', on reading the specification, is unable to make the invention work.

4. Unfairly based, meaning that the complete specification contains subject matter which was not contained in, or foreshadowed adequately by, the provisional specification.

5. Unfairly obtained, which is a polite way of implying that the applicant was not the true and first inventor, but obtained the invention by fair or foul means.

If the opposition period expires without any attack against the specification having been made, the application proceeds to being granted and sealed. But even then the troubles of the owner of the patent are not over yet; first of all there is a period of 12 months during which application may be made to the Comptroller (the head of the Patent Office) for revocation of the patent on substantially the same grounds as during the opposition period. And thereafter the patent may still be revoked on the ground that it is not being worked, or because of inutility.

Furthermore, the life of the patent is dependent on the payment of renewal fees which increase in severity as the patent gets older. These renewal fees ascend from £30 for the fifth year to £84 for the sixteenth and final year. The total amount of official fees payable during the normal life of a British patent is £696, provided that no complications, such as opposition, revocation and the like occur. These fees are those payable to the Patent Office, and since it is advisable, when dealing with patents, to utilise the services of a qualified British patent agent, there will be additional fees to be paid to the agent. These professional fees are broadly based on a scale laid down by the Chartered Institute of Patent Agents.

The history of a British patent may be tabulated as in *Table 6.1*.

Table 6.1

	Maximum time interval (months)	Official fee (£)
Provisional application (optional)	12	1
Complete specification		48
Examination		
Acceptance	approx. 1½	
Publication		
Opposition	3	
Grant and seal	approx. 1½	17
Revocation	12	
Renewal fee in respect of the 5th year		30
6th year		32
7th year		34
8th year		38
9th year		42
10th year		48
11th year		52
12th year		60
13th year		64
14th year		70
15th year		76
16th year		84

The life of a British patent of 16 years is reckoned from the date of filing the complete specification and may under exceptional circumstances be extended, particularly if it can be shown that the invention was so much ahead of its time that the inventor was prevented from obtaining sufficient reward for his work.

Information aspects

Apart from the purely legal protection and rights afforded by a patent, such as the monopoly extending for a number of years over the entire area of the United Kingdom of Great Britain and Northern Ireland and the Isle of Man, patents can provide an extremely useful source of technical information whose value is so often underestimated. The reasons for such a neglect may be found in the previously mentioned excessive use of the legalistic language of 'patentese' and the uninspiring appearance of the specifications themselves; but it seems that by far the strongest factor is the ignorance among scientists, technologists and information scientists of the value of information contained in the patent literature.

The magnitude of the problem can perhaps best be appreciated by realising that approximately half a million patent specifications are published each year, of which less than half are published in the English language—in the USA 74 182 patents were published in 1973 and 40 600 in the UK. A recent study[1] has shown that out of 1000 patents selected at random from three separate years of publication, only about 5·8% had their contents published in other forms of literature. So how much valuable material is there hidden in the very large remainder?

The reason just suggested that patents may be the only place of disclosure of information is, however, but one for urging a wider use of patent literature. The others are that in many cases patents disclose an important subject much sooner than other sources. From the numerous examples one may quote:

	Date of published patent	Date of first disclosure in other forms of literature
Hollerith (punched card)	1889	1914
Baird (television)	1923	1928
Whittle (jet engine)	1936	1946
Morrogh (ductile iron)	1939	1947

Furthermore many patent specifications contain much more detailed information than other types of literature. Some patent specifications are almost textbooks in themselves; for example, the three-volume British patent 1 108 800 contains a wealth of details of the IBM 360 computer system.

But the methods of extracting the information from this flood of patent literature are none too easy. There are abstracting services devoted exclusively to patents which permit of both a current awareness service so as to enable the reader to keep abreast with developments as disclosed in the patent literature, and also, to some extent, retrospective searching, so that earlier work can be retrieved. In the UK the Patent Office produces an 'Abridgement service' which provides abstracts of recent British specifications in a non-patentese language, arranged numerically in the 25 groupings listed in *Table 6.2.*

Table 6.2

Classification Key Unit	Abridgment Volume Covering Division(s)	Subjects covered
1	A1–3	Agriculture; animal husbandry; food; tobacco; apparel; footwear; jewellery
2	A4	Furniture; household articles
3	A5–6	Medicine; surgery; pesticides; fire-fighting; entertainment
4	B1–2	Physical and chemical apparatus and processes; crushing; coating; separating; spraying
5	B3	Metal-working
6	B4–5	Cutting; hand-tools; radioactive material; working non-metals; presses
7	B6	Stationery; printing; writing; decorating
8	B7	Transport
9	B8	Conveying; packing; load-handling; hoisting; storing
10	C1	Inorganic chemistry; glass; fertilisers; explosives
11	C2	Organic chemistry
12	C3	Macromolecular compounds
13	C4–5	Dyes; paints; miscellaneous compositions; fats; oils; waxes; petroleum; gas manufacture
14	C6–7	Sugar; skins; microbiology; beverages; metallurgy; electrolysis
15	D1–2	Textiles; sewing; ropes; paper
16	E1–2	Civil engineering; building; fastenings; operating doors, etc.
17	F1	Prime movers; pumps
18	F2	Machine elements
19	F3–4	Armaments; projectiles; heating; cooling; drying; lighting
20	G1	Measuring; testing
21	G2–3	Optics; photography; controlling; timing
22	G4–5	Calculating; counting; checking; signalling; data handling; advertising; education; music; recording; nucleonics
23	H1	Electric circuit elements; magnets
24	H2	Electric power
25	H3–5	Electronic circuits; radio receivers; telecommunications; miscellaneous electrical techniques

These abridgements are prepared by the examiner after he has read the specification, and are usually accompanied by drawings. After a group of 25 000 specifications has been published, name and subject indexes are prepared to facilitate retrospective searching. The British abridgements have been published since 1855, so that the technological developments during the last century can be followed and searched with comparative ease, provided that the subject arrangements and classification are grasped.

The Patent Office is also prepared to provide an inquirer with file lists, which are tabulations of the serial numbers of all specifications which have been allotted a classifying or indexing code, thereby giving a ready means of identifying patents which disclose a particular subject matter. Several types of file list can be bought from as little as 60p per list going back to specifications published in 1911. In some classes it is even possible to obtain combinations of up to 25 code numbers.

For specifications relating to coatings, titania, lubricants, alloys and stimulated emission devices 80-column direct-coded punched card systems can be bought which are updated every 2 months.

The increasing use of computer systems for information retrieval has led at least one organisation, Derwent Publications Ltd, to launch in 1974 an ambitious scheme of providing data on all patents from 24 countries by its World Patent Index. This computer-generated phototypeset service covering all subjects includes fully indexed information on over 12 000 patents published each week.

There are other literature services which treat patents just like any other form of literature by making abstracts of them, indexing them and retrieving information from them. Among the foremost of these services is *Chemical Abstracts*, whose comprehensive coverage of the chemical literature started in 1907 and has been steadily improved ever since. At present, patents from 25 countries are scanned regularly and abstracted by experts who have been carefully trained to write long and informative abstracts which are often accompanied by numerous examples. From the standpoint of providing a current awareness service, *Chemical Abstracts* may occasionally be rather slow, with the result that in some cases the opposition period may have expired before the abstract of the specification has been published. On the other hand, the well-known depth of indexing employed by *Chemical Abstracts* considerably facilitates retrospective searching through earlier issues. The subject field of this service, though nominally devoted to chemistry, extends considerably beyond the strict definition and frequently topics of other disciplines are covered, such as physical chemistry, optical test methods, thermoluminescence, etc.

While the general literature of physics is well served *inter alia* by the various INSPEC publications, especially *Physics Abstracts* and *Current Papers in Physics*, their coverage of the patent literature is not up to the same standard as that of the *Chemical Abstracts* services. Recent proposals such as the Patent Associated Literature Service are designed to remedy this deficiency. The major French-language abstract service, *Bulletin Signalétique*, like its German sister publication, *Physikalische Berichte*, seems to ignore the existence of patents altogether, while, on the other hand, the Russian-language *Referativnyĭ Zhurnal* contains abstracts of many patents, especially of Eastern European countries.

The future of patents

As mentioned earlier, patents have a legal significance which is, however, restricted usually to the territory of the issuing country. Thus, a British patent specification enjoys legal protection only within the confines of the United Kingdom of Great Britain and Northern Ireland and the Isle of Man. Even the Channel Isles have their own patent systems; and in order for an inventor to obtain extensive protection for his intellectual property, he has to apply for patents in each country where he wishes to obtain such protection. Attempts have been made for nearly a century to establish a system whereby a ·patent extending beyond one country may be obtained by filing a single application. The concept of a 'world patent' is old indeed, but has been very slow in materialising. Several political moves in recent years have, however, accelerated developments, so that at present there are at least three moves towards supranational patent systems. These are:

1. The EEC patent, which would result in the issue of one patent to be valid in all nine Member States of the European Economic Community (Belgium, Denmark, France, Germany, Ireland, Italy, Luxembourg, Netherlands and the United Kingdom) with a central patent office situated in Munich.

2. The European patent system under the aegis of the Council of Europe, in which 21 European states will collaborate to have a parallel system of national patents (as at present) and a supranational European patent valid in those countries the applicant selects; in this system the EEC countries will count as one 'patent country'.

3. The Patent Co-operation Treaty, which was signed in Washington in 1970. The proposals of this treaty provide essentially for two phases: in the first an applicant can file an international application which can become individual applications in those countries he cares

to designate. An international novelty search will be carried out to prescribed standards in one of a number of approved search centres, which will probably be at Washington, Munich, Moscow and Tokyo and at the International Patent Institute at The Hague. Under Phase II of this treaty the report of the international novelty search will be accepted by the national patent offices as a basis for granting or refusing a patent.

In each of these three proposals there is a significant absence of any reference to the British Patent Office, which will in effect wither away to become merely a forwarding and recording centre for applications originating in the UK.

TRANSLATIONS

The importance of the barrier caused by the existence of the many foreign languages used in the communication of scientific and technical information is not always fully appreciated. Even the very fact that there are languages other than one's native tongue and that there are altogether more than 3000 languages—leaving out many more dialects —is frequently overlooked or conveniently forgotten. It is, of course, also true that most of these languages do not make a significant contribution to the vast amount of literature which is produced and that perhaps only some ten or so languages play an important part. Yet the size of the language gap is so great that it may be dangerous to ignore it or to minimise its significance.

One of the dangers of neglecting foreign-language material— indeed, published material as such—is that such an action could lead to dangerous and costly wastage of time and money by necessitating the duplication of work which had previously been done and described in the literature. There are numerous examples of such occurrences.

Although it is difficult to be very precise in assessing the quantitative nature of the language problem, recent analyses of the language distribution in the literature of science and technology—compared with earlier investigations—show some startling figures. In 1957 the first serious attempt was made to obtain an accurate picture of language distribution when Holmstrom and Lloyd[2] studied 1000 scientific periodicals selected at random and found that 43·6% were published in English, 14·4% in German, 12·6% in French and 8·1% in Russian. These figures were accepted without much question for almost a decade. In 1965 Tybulewicz and Liebesny[3] conducted a further study on a slightly different basis by analysing the language of the original articles abstracted in a variety of abstract journals. This investigation showed three important points: (1) that English is still the most widely used language in science and technology, accounting for more than half the world's output; (2) that the second most important language is Russian, representing some 20% of the literature; (3) that half of the world's scientists cannot read half of the world's scientific literature, purely because it is in languages which they do not understand.

To put these percentages into absolute figures it means that about 800 000 scientific and technical articles are published annually in English and some 300 000 in Russian. The reasons for the preponderance of Russian over other foreign languages in the disciplines of science and technology are numerous and are obviously interrelated, but include the facts that:

1. The quantity of Russian literature *per se* has increased considerably, owing to the rapid rate of technological advance prevailing in the Soviet Union, which rate is probably greater than that in many Western countries.

2. The means of reporting Russian literature have improved greatly since 1957, when the launching of Sputnik I took place. This event prompted the setting-up of cover-to-cover translations which have most definitely multiplied the coverage of the Russian output of scientific publications.

3. The number of Russian abstracting services (the *Referativnye Zhurnaly*), which scan comprehensively the less accessible Soviet periodicals, has grown.

4. Russian authors appear more willing to publish the results of their work.

One aspect which may have considerable significance in years to come is that the Chinese contribution, although still small, has risen by a factor of about 100 compared with 1961.

In the field of physics[4] the present picture shows that the English, Russian, French and German contributions together make up some 97% of the total; but it may be useful to remember the old riddle in the translation industry: 'What is the difference between an optimist and a pessimist?' Answer: 'The optimist learns Russian, while the pessimist learns Chinese.'

To make this large amount of foreign language material accessible and comprehensible to the potential user—the research scientist, engineer, manager, patent officer, etc.—it is essential to bridge the language gap in an efficient and effective manner, since the user is rarely capable of properly understanding an original article in a foreign language; he will almost invariably demand a rendering of it into his own language. But it is generally rare to find people who are equipped to deal with much more than one or two foreign languages.

Hence, it becomes necessary to resort to translations and to the people capable of producing them, viz. the translators.

Many years ago the famous French writer Jean Baptiste Racine (1639–99) defined the requirements of a translation as follows: 'Les traductions sont comme les femmes. Si elles sont belles, elles ne sont pas fidèles, et si elles sont fidèles, elles ne sont pas belles.' This quotation crystallises the difficulty facing every translator: should he produce a really accurate translation which may read badly, or should he produce a text which reads as if it was written originally in that language, thereby, however, losing some of the flavour and accuracy of the original version? In the scientific and technical translation field the translator must resolve that problem, probably in favour of the 'accurate rendering' alternative; but above all he must possess two very important qualifications: (1) a thorough knowledge of the two languages, the one he is translating from and the one he is translating into (sometimes called the 'input' and 'target' languages, respectively), and (2) a mastery of the subject matter of the text.

As far as the first requirement is concerned, it is normal practice in linguistic circles to require of a translator to translate only into his mother tongue or language of habitual use. With regard to the second requirement, it is surely unfair to the user and the translator if the latter is asked to deal with an article on, say, electron microscopy if his expertise lies in agricultural machinery. Clearly the special terminology of the subject will be unfamiliar to him, probably in both languages, unless he makes a special study of it, and he will therefore be unable to fully understand the text and, hence, to translate it correctly.

For a translator to be efficient at any level of proficiency it is, how-ever, equally important to have available really reliable tools, i.e. dictionaries. Although such works are usually classed as reference works and thereby given a status of infallibility, they rarely deserve it. It must always be remembered that dictionaries are compiled by human beings, who, alas, are frail and fallible. Hence, the usefulness of dictionaries which are compiled by such beings is often adversely affected by inaccuracies, omissions, distortions, typographical errors, etc. Some compilers of dictionaries appear to rely on people with mainly linguistic qualifications and thereby overlook the previously mentioned importance of specialised subject knowledge.

For example, the publishers of a well-known French-English dictionary have been known to use schoolmasters to check the entries; it is reasonable to assume that few of these teachers have ever been inside a foundry and would therefore be utterly unaware of the fact that the French word 'fonte' has at least three different meanings in foundry technology alone. As a result the translation of that word into English—when based on the definitions in that dictionary—may be quite quaint and even inaccurate. Perhaps Alice's friend Humpty Dumpty in *Through the looking-glass* is the patron saint of all dictionary compilers by virtue of his motto: 'When I use a word it means just what I choose it to mean—neither more nor less.'

It should be noted that only rarely does any dictionary give examples of context to illustrate various shades of meaning. Many others consider it sufficient to provide a few symbols such as intertwined anchors, for example, to denote the whole wide world of marine terminology.

Since translating is a skilled task carried out by professionals, the cost of translations and the time factors involved in producing a translation must be considered. The costs of a translation depend mainly on three factors: (1) the combination of languages, (2) the complexity of the text, and (3) the urgency. As far as the language is concerned, it will be found that the more common languages, such as German, French, Italian and Spanish, are less expensive than the more difficult ones, such as Russian or Japanese. The basis on which the costs of translations are worked out is 1000 words of original text, and the Institute of Linguists—the professional body of translators, interpreters, language teachers, etc., which was established in 1910— has issued a list of recommended *minimum* charges for translations from and into various language groups. Since, however, most texts to be translated are of a complicated nature, the charges applicable to them are considerably higher than those shown in that tariff and normally range from about £10–15 per 1000 words from German, French, etc., into English to £20–25 from Japanese per 1000 characters. Translations out of English command as a rule a 50% surcharge on those rates.

Since it is obviously more difficult to translate a text on, say, diffraction optics than a simple letter from a customer saying, 'Dear Sir, unless payment of our invoice of 15th ult. is forthcoming immediately...', the charges for the more complex text will be higher. Similarly, a translation which has to be completed in a very short space of time and will thus mean working during the so-called unsocial hours will cost more than one done at a normal pace. Rough estimates of workloads of translators show that the normal output is approximately 2000–5000 words per day, i.e. between 5 and 12 pages of this book.

Whenever translation of a complex nature and/or extreme urgency is required, it is, however, advisable to make use of the more varied and flexible services of a translation agency, since such organisations have access to a large number of individual translators, either on their permanent staff or retained by them so that a long text can, for example, be divided up and then edited centrally for style and uniformity. Furthermore, if translations are commissioned from English into a foreign tongue, then the agencies may have links with similar institutions in other countries so that the local usage of that foreign language can be utilised to the full.

In the UK there are two large indexes of translators: the one maintained by Aslib (3 Belgrave Square, London), which is carefully co-ordinated to enable suitable matchings of language and subject competence to be realised; and the one published by the Translators Guild of the Institute of Linguists (91 Newington Causeway, London), which is updated by quarterly supplements.

Before a translation is commissioned, it is advisable to ascertain whether a translation has not already been made of that text. In addition to the enormous annual compilation by Unesco of the *Index translationum*, which refers to published translations, mainly of books, there is the *Commonwealth index of unpublished scientific and technical translations*, whose centre is located at Aslib (for address, see above), where bibliographical references to about 250 000 translations are recorded together with indications where the translations may be obtained. A steady influx of notifications of completed translations is incorporated from many sources, including the European Translations Centre (101 Doelenstraat, Delft).

From the inquiries which Aslib receives for locations of translations, about one in eight can be answered positively with the reply that a translation of that particular item has been made and that a copy of the translation is available. The British Lending Library (formerly the National Lending Library for Science and Technology) at Boston Spa, Yorkshire, also maintains such an index with specific reference to Russian and other Eastern European languages for which group of languages another helpful medium has been provided. Since the late 1950s, when the first Russian Sputnik made its appearance in the sky and the Western world realised that up till then they had ignored the Russian literature, the more important Russian periodicals in the fields of science and technology have been translated from cover to cover into a more accessible Western language such as English or German. Although these cover-to-cover services are slow and expensive, they provide a very valuable tool for bridging the language gap.

The abstracting services are also extremely useful aids in this respect, since the better services cover the foreign-language literature in considerable depth and provide an indication—in English—of the contents of periodicals, books, reports, patents, standards, etc.

The multitude of languages mentioned above creates a considerable barrier to mutual understanding and therefore numerous attempts have been made in the past to overcome it by creating a universal language. Most of these endeavours have not been crowned with success, but reference should be made to the following:

Latin was used by scientists (such as Bacon, Newton, Descartes) for a very long time as a universally understood medium, but its complex structure and the fact that it is spoken by only very few people have caused it to be used very rarely in modern times.

Volapük, the first artificial language, was invented in 1880 by Johann M. Schleyer, a priest of Constance, Baden. Its vocabulary is mainly based on English, though with a greatly simplified grammar. The word means 'world speech'.

Esperanto, the best known of the synthetic languages, was devised by a Russian physician, Dr L. L. Zamenhof, in 1887. Although it has more than 1250 local societies and national affiliates in 21 countries, it is always a second language to its speakers. Its use in science and technology is very limited.

Ido was developed from Esperanto in 1907 (its name is the Esperanto word for 'offspring'), but its growth has been rather slow and its influence is almost insignificant.

Basic English. This simplified form of English with a vocabulary of 800 words was introduced during World War II, mainly to help foreigners to learn English. Some scientific texts were produced in this language, but they were not considered to be very successful.

With the present preponderant position of English in many fields of scientific and technological documentation, it may be expected that English will become the major international means of communication in the future. Many of the developing countries have accepted this probability and their scientists now tend to publish their findings in English, perhaps thereby to gain a wider readership for their contributions.

While the foregoing has been mainly concerned with the intellectual efforts of humans in translating texts which are sometimes of doubtful quality, one should not forget the efforts that have been made especially since the end of World War II in harnessing the electronic computer for such operations. However, the high hopes placed on mechanical or machine translation—as it became known—have not been materialised. It all sounded so easy that the action of translating, say, CAT from English into, for example, French, means merely a mathematical equation which has to be expressed in binary digits, viz.

C	A	T	=	C	H	A	T	
3rd	1st	20th		3rd	8th	1st	20th	letter of
								alphabet, or
00011	00001	10100		00011	01000	00001	10100	in binary code

and then be stored in the capacious memory unit of the computer. But the complexities of semantics (e.g. CAT can be either a feline quadruped or a type of whip or a burglar or an abbreviation for either a catalytic cracker or a college of advanced technology), syntax (the inversion, as in French, of nouns and adjectives such as 'moulin rouge' or 'carte blanche' or the German predilection for placing the verb at the very end of a long sentence), grammar (the lack of auxiliary verbs or articles in Russian), etc., proved to be too difficult, and nowadays mechanical translation may be considered moribund.

CONCLUSION

There is a widely held belief that scientific and technical literature in a foreign language is in a different category from English-language material, especially as far as quality is concerned. This erroneous assumption has been disproved many times.

The size of the language problem is truly formidable in terms of quantity, intellectual effort (of translating, editing, etc.) and finance. Yet, in spite of its magnitude, the problem of the language barrier can be overcome by various means with differing degrees of effectiveness. All these steps require at some stage or other the intervention of a yet unreplaceable human element—the skilled professional linguist. Any inferior substitute—such as the person having spent 3 weeks' holidays in Spain or the French au pair girl with no technical knowledge—must be firmly rejected. The real professional translator will, however, provide invaluable assistance in bringing together different worlds of thought and development in order to advance the progress of knowledge. The effectiveness of his contribution depends materially on various tools and systems, many of which are still capable of considerable improvement.

REFERENCES

1 OSTI Report. 'The scientific and technical information contained in patent specification—the extent and time factors of its publication in other forms of literature.' July 1973
2 *Scientific and technical translating and other aspects of the language problem.* p. 15. Paris, Unesco, 1957. Also *Information Scientist*, **8**, 165–177 (1974)
3 Tybulewicz, A. and Liebesny, F. 'The relative importance of various languages in scientific and technical literature' (*Inc. Linguist*, *4* (1), 12–13, Jan. 1965)
4 Tybulewicz, A. 'Languages used in physics papers' (*Physics Bull.*, (120), 19–20, Jan. 1969)
See also
Liebesny, F. (ed.). *Mainly on Patents*. Butterworths, 1973

7

History of physics

Alison R. Dorling

The great rise in the number of history of science courses in universities and colleges over the past two decades has to a considerable extent been made possible by the work of the reprint companies. The existence of reprints means that the person interested in tracing the rise of some branch of physics, and without access either to large national collections or to the older university libraries, is no longer restricted to popular and often misleading, if not actually woefully inaccurate, histories. The original developers of particular theories may not always have expressed themselves with the clarity or rigour of the tidied-up version in a modern textbook, but the motivation for the work is often much more clearly to be seen.

Foremost among the reprint firms in this field is Dover Publications, handled in the UK by Constable, whose editions are almost all soft-backed, and so reasonably priced. The one serious flaw that can be attached to them in general is that they rarely give adequate information about the original publication details. Another inexpensive reprint series of value is Pergamon's 'Selected Readings in Physics' series. Each volume collects together a group of historically important papers in a particular branch of physics and contains an introduction, usually by a physicist working in the field, to the history of the subject and the significance of the papers to contemporary physics. Several examples are discussed in more detail later in the chapter.

Of course, there are other reprint series in the history of science, such as the 'Cass Library of Science Classics', published in London; the 'Collection History of Science', published in Brussels by Éditions Culture et Civilisation; and Johnson's 'The Sources of Science' in New York, containing, for example, Priestley's *The history and*

83

present state of electricity (1767; Johnson Reprint Corp., 1966)*; as well as individual works reprinted by various companies: for example, Boltzmann's *Populäre Schriften* (1905; 'Journalfranz' Arnulf Liebing, 1972), and others which are mentioned later in the chapter. These tend to be aimed at the library market, and to be correspondingly expensive.

The following pages briefly discuss the history of the various branches of physics and the most readily available source materials. After each work discussed, the original date of publication and the publisher and date of a modern edition are given. Further details can be found in the bibliography at the end of the chapter.

MECHANICS AND RELATIVITY

Galileo's work on statics and falling bodies is primarily to be found in the *Dialogues concerning two new sciences* (1638; Dover, 1951), in which Galileo, in the person of Salviati, expounds his subject in the face of the objections of an Aristotelian, Simplicio, with the mediation of a representative of the common-sense everyday point of view in the person of Sagredo (who, of course, eventually decides in favour of Salviati).

Most of Descartes' work in physics is to be found in the *Principes de la philosophie* (1664; Vrin, 1964). There is not a complete translation of this into English, and the sections which have been translated have been intended mainly for philosophers and have not included, for example, his attempts to formulate laws of motion.

Newton's *Principia* (1687), on the other hand, is available in many versions. The only paperback one is Motte's original English translation, revised in 1934 by Cajori (University of California Press, 1962). The revision produced a less satisfactory translation in some crucial places, and some very uneven editorial notes. The most recent, and most scholarly, edition is that edited by I. Bernard Cohen (CUP, 1971–), of which the first volume consists entirely of editorial matter, discussing, for example, the changes made by Newton between successive editions. Westfall's *Force in Newton's physics* (Macdonald, 1971) and Dugas' *Mechanics in the seventeenth century* (Éditions du Griffon, 1958) both discuss the theories of the period in some detail.

Early work in fluid and continuum mechanics is discussed in long editorial essays by Truesdell to Volumes 11/2, 12 and 13 of Euler's collected works, *Leonhardi Euleri opera omnia* (Füssli, 1960, 1954,

* Full bibliographical entries for all items described are given at the end of this chapter.

1956). These consider fluid mechanics, the mechanics of elastic bodies, and sound up to the year 1788 (a separate essay is devoted to Euler's *Treatise on fluid mechanics*). In that year Lagrange's *Méchanique analytique* (1788; Blanchard, 1965) was published.

The work of Newton's successors in the eighteenth century, and in particular of Laplace, was analysed by Todhunter in his *History of the mathematical theories of attraction and the figure of the earth* (1873; Dover, 1962). Mach's *The Science of Mechanics* (1883; Open Court Publishing Co., 1960) is somewhat deficient as history but is nevertheless important—for example, in undermining the concepts of absolute space and time. Todhunter was also responsible, together with Pearson, for a *History of the theory of elasticity and the strength of materials* (1886–93; 3 vols., Dover, 1960). Later fluid mechanics is only covered, in a very superficial manner, in Tokaty's *History and philosophy of fluid mechanics* (Foulis, 1971).

In sound the principal classic is Rayleigh's *Theory of sound* (1877; Dover, 1956), since Helmholtz's work belongs more properly to the history of physiology than to the history of physics. More recent developments in continuum mechanics are outlined in Truesdell's *Six lectures on natural philosophy* (Springer, 1966). A textbook with a semi-historical approach is Yourgrau and Mandelstam's *Variational principles of dynamics and quantum theory* (Pitman, 1960).

The development of special and general relativity can be followed through the papers of Einstein, Lorentz and Minkowski in *The principle of relativity* (1922; Dover, 1952). Kilmister's *Special theory of relativity* (Pergamon, 1970) reproduces other important papers, in particular those of Michelson and Poincaré. Of the latter, Whittaker argued, in the second volume of his *History of the theories of aether and electricity* (Nelson, 1953), that relativity owed more to Lorentz and to Poincaré than to Einstein. One of the clearest presentations of the general theory is still Eddington's *Mathematical theory of relativity* (CUP, 1924).

LIGHT, ELECTRICITY AND MAGNETISM

The first major scientific work by an Englishman was Gilbert's *De magnete* (1600; Dover, 1958), important for its experimental contributions to electrostatics, as well as for those to magnetism. After describing his experimental work, Gilbert introduced the idea that the earth is a magnet, and, having no preconceptions about scientific method, 'explained' the motion of the solar system magnetically on the basis of some unsatisfactory analogies. Thereafter, although the art of navigation developed, no great advances in magnetism were made for nearly two centuries, until the work of Coulomb.

In the meantime Huygens and Newton were introducing their theories of optics. The theory introduced by Huygens in his *Treatise on light* (1690; Dover, 1962) is not a wave theory in the modern sense, since it does not introduce wavelengh, and his method of constructing wave-fronts is difficult to reconcile with his own underlying mechanical model. Newton's *Opticks* (1704; Dover, 1952) contains, as well as his account of his experiments and theory, his speculations on a wide range of physical (and theological) questions. Good analyses of the scientific arguments and claims of these theories is given in Sabra's *Theories of light from Descartes to Newton* (Oldbourne, 1967).

The contributions of Young, Arago and Fresnel to the development of optics in the early nineteenth century have, unfortunately, not been reprinted. A brief account of their work and of the conflict with corpuscular theory (a minor affair compared with the conflict between Newtonians and Leibnizians over the calculus) is given towards the end of Ronchi's history *The nature of light* (Heinemann, 1970). Thomas Young worked in a variety of fields: physiological optics, physics, medicine, linguistics (the decipherment of the Rosetta Stone); but his importance has sometimes been overrated. There is a tendency to find in experts on any of these fields the sentiment that of course Young was important, but his contributions were primarily in other fields.

The developments in optics in the latter half of the nineteenth century were more technical in character. Rayleigh's contributions can be found in his *Scientific papers* (3 vols., Dover, 1964). The state of optical theory at the turn of the century can be seen from Drude's *Theory of optics* (1900; Dover, 1959). Of more recent texts, Ditchburn's *Optics* (Blackie, 1963) is semi-historical in its presentation. Mach's posthumous *Principles of physical optics* (1921; Dover, n.d.) concentrates on the earlier history, and is disappointing compared with his work on mechanics.

Apart from the development of the wave theory and advances in the mathematical theory of optics, the other great development in the nineteenth century was in spectroscopy. This has recently been described in McGucken's *Nineteenth century spectroscopy* (Johns Hopkins Press, 1969), but this, unfortunately, does not give the modern explanations of the phenomena he discusses, and indeed stops short of the most interesting developments in the subject.

The development of electricity and magnetism can be followed more easily than that of optics. Franklin's *Experiments and observations on electricity* (1751; Harvard University Press, 1941) consists of letters to Peter Collinson, a botanist and Fellow of the Royal Society who communicated Franklin's work to the Society. (Franklin was annoyed that they were not published in the *Transactions*, and clearly regarded them as public in character.) The Harvard edition contains an extensive introduction by I. Bernard Cohen to Franklin's life and to the context and significance of his work.

A collection of Coulomb's *Mémoires* from 1777 to 1806 was published by Gauthier-Villars in 1884, and remained on their list until a few years ago. Ampère's *Théorie mathématique des phénomènes électrodynamiques* (1827; Blanchard, 1958) has not been translated *in toto*, and the extracts given in Tricker's *Early electrodynamics* (Pergamon, 1965) have been badly mauled in translation. Ampère's book is of interest for its discussion of scientific method. Tricker's book also contains Oersted's main paper, and extracts from Biot and Savart, and Grassmann.

Dover has reprinted both Faraday's *Experimental researches in electricity* (1839–55; Dover, 1965) and Maxwell's *Treatise on electricity and magnetism* (1873; Dover, 1954). The Faraday and *The scientific papers of James Clerk Maxwell* (Dover, 1965) are only available in cloth-bound editions. Faraday's style is long-winded, and an introduction to his work may be made by way of the selections in Tricker's *The contributions of Faraday and Maxwell to electrical science* (Pergamon, 1967). Although Maxwell's interpretation of his theory differs in many respects from the modern one (in particular, a satisfactory justification of the displacement current term cannot be found in his works), his derivation of some of his results still remains the accepted one. For example, Lamb, in describing certain aspects of vortex motion in his *Hydrodynamics* (CUP, 1932), refers the reader to Maxwell's derivation of the equations for the analogous electromagnetic situation in the *Treatise*.

However, the equations we now call Maxwell's Equations were first put into their present form by Hertz and Heaviside, and can be found in Hertz's *Electric waves* (collected papers published in 1892 and dating from the previous 5 years; Dover, 1962). Another important classic of the subject is Lorentz's *Theory of electrons* (1909; Dover, 1952). The standard history of electromagnetism is the first volume of Whittaker's *History of the theories of aether and electricity* (1910; Nelson, 1951)—more for reference than a book to be read straight through. A very good critical account of the development of electromagnetic theory, from an unorthodox point of view, is given in O'Rahilly's *Electromagnetic theory* (published in 1938 as *Electromagnetics*; Dover, 1965). A rather broader field, ranging from the Greeks to quantum mechanics and dealing with such problems as action at a distance, is covered in Hesse's *Forces and fields* (Nelson,

1961). The more recent history, the development of relativistic electromagnetic theory, is covered in the historical sections of Rohrlich's *Classical charged particles* (Addison-Wesley, 1965).

HEAT, THERMODYNAMICS AND STATISTICAL MECHANICS

The first attempt to erect a theory of heat was made by Fourier in his *Analytical theory of heat* (1822; Dover, 1955), on the basis of his study of conduction. (Anyone interested in scientific method might like to start further back with Bacon's *Novum Organum* (1620; Liberal Arts Press, 1960), the earliest book on the subject, which contains as an example of his method a 'deduction' of the kinetic nature of heat.)

Carnot introduced the Second Law of Thermodynamics in his *Reflections on the motive power of fire* (1824; Dover, 1960); the First Law, although taken for granted already by Leibniz in the form of the non-existence of a perpetual motion machine, was not incorporated into theory and used to derive physical results until the work of Helmholtz and others in the middle of the nineteenth century. The Dover edition of Carnot's book contains also the paper by Clapeyron which made Carnot's work known to the scientific community of the day, and a paper by Clausius reconciling the First and Second Laws.

Much important work on the development of statistical mechanics has been done by Brush, and in his two volumes entitled *Kinetic theory* (Pergamon, 1965 and 1966) he reprints a wide selection of material: from Boyle to Clausius in the first volume ('The nature of gases and of heat'), which includes the work of Helmholtz and others referred to above, and from Maxwell to Boltzmann in the second ('Irreversible processes'). His introductory material occupies a much smaller proportion of the volumes than is common in this series, but constitutes a reliable brief guide to the subject. Brush has also translated and edited Boltzmann's *Lectures on gas theory* (1896–98; University of California Press, 1964).

An entirely new approach was introduced by Gibbs in his *Elementary principles in statistical mechanics* (1902; Dover, 1960); his papers on thermodynamics are reprinted in the first volume of *The scientific papers of J. Willard Gibbs* (Dover, 1961). Both Boltzmann's and Gibbs's approaches were critically discussed in the Ehrenfests' survey *The conceptual foundations of the statistical approach in mechanics* (1912; Cornell University Press, 1959).

The Third Law is due to Nernst, whose *New heat theorem* (Dover, 1969) describing his contribution over the previous 20 years was originally published in 1918; the existence, as opposed to the unattainability, of absolute zero had been known from the end of the eighteenth century. Mendelssohn's *The quest for absolute zero* (Weidenfeld and Nicolson, 1966) is a popular account of the experimental developments in the twentieth century, made all the more interesting for its author's participation in those events.

The recent history of the subject, particularly of the statistical interpretation of the Second Law, can be found in Jancel's *Foundations of classical and quantum statistical mechanics* (Pergamon, 1969).

QUANTUM THEORY

Early quantum theory, prior to the introduction of wave mechanics by Heisenberg and Schrödinger, was able to give many atomic phenomena at least a qualitative explanation. The selection of papers edited by Ter Haar, *The old quantum theory* (Pergamon, 1967), contains work by Planck, Einstein, Rutherford, Bohr and Pauli. A good account of what had been achieved shortly prior to modern quantum theory is contained in Lorentz's *Problems of modern physics* (1927; Dover, 1967), which reproduces lectures he gave in 1922. The second volume of Whittaker's *History*, referred to previously, has now largely been superseded by Jammer's reliable and comprehensive *The conceptual development of quantum mechanics* (McGraw-Hill, 1966), but occasionally explains an argument with more mathematical detail. Jammer's book takes the development of quantum mechanics up to 1926.

The beginnings of wave mechanics are already to be found in de Broglie's thesis *Recherches sur la théorie des quanta* (1924; Masson, 1963). A small part of this is translated in Ludwig's *Wave mechanics* (Pergamon, 1968), together with more extensive extracts from Schrödinger, and some work by Heisenberg, Born and Jordan. Overlapping to some extent with Ludwig is Van der Waerden's collection *Sources of quantum mechanics* (Dover, 1968), which concentrates on the development of the matrix approach to quantum mechanics. Lectures which Heisenberg gave in 1929 were published as *The physical principles of the quantum theory* (1930; Dover, 1949?). These required a very advanced knowledge of quantum theory when they were given, and present a more sophisticated discussion of wave-particle duality than he had given previously.

The only comprehensive and consistent account of Schrödinger's classical wave theory, which he himself never developed in any detail, is given in an early chapter in the second volume of Tomonaga's *Quantum mechanics* (North-Holland, 1966). The first volume is concerned with the old quantum theory, and explains all the mathematical and physical arguments in detail. The presentation is often close to that of the original papers, but, unfortunately, without references.

More attention has been devoted to the development of quantum theory than to any other aspect of the history of physics, and anyone interested in the subject should be aware of the project carried out by Kuhn and his colleagues under the auspices of the American Philosophical Society and the American Physical Society, and reported in *Sources for the history of quantum physics* (American Philosophical Society, 1967). This provides a detailed inventory of all the surviving notebooks, letters and other unpublished material of physicists concerned in the development, and of the transcripts of interviews with more than 90 such people obtained by members of the project, and the location of this material.

PARTICLE PHYSICS

Despite the comparative recency of the growth of particle physics, a number of collections of papers illustrating its development are available. *The world of the atom* (2 vols., Basic Books, 1966), edited by Boorse and Motz, is a comprehensive collection ranging from Lucretius to quarks, and containing much material of relevance to kinetic theory and quantum theory, as well as Nobel Prize lectures or primary papers by the main contributors to particle physics. *Foundations of nuclear physics* (Dover, 1949), edited by Beyer, is an early collection containing 13 papers and an extensive but badly organised bibliography of relevant material. Schwinger's *Selected papers on quantum electrodynamics* (Dover, 1958) starts with a paper of Dirac's in 1927 and takes developments up to the mid-1950s. Kabir's collection *The development of weak interaction theory* (Gordon and Breach, 1963) is self-explanatory. Strong interaction theory in the period 1961–64, during which SU(3) established itself as a central organising principle of particle physics, is dealt with in Gell-Mann's *Eightfold way* (Benjamin, 1964). A direct sequel to this, covering the larger symmetries, and the question of the consistency of SU(6) with relativity, is Dyson's *Symmetry groups in nuclear and particle physics* (Benjamin, 1966).

The preceding sections have merely sketched the materials available for the history of the main branches of physics. There are two books available which can be consulted for more comprehensive surveys.

The first is Brush's *Resources for the history of physics* (University Press of New England, 1972), published for the International Working Seminar on the Role of History of Physics in Physics Education, and intended primarily to make teachers of the history of physics aware of all the relevant material in print. It has good international coverage, providing, for example, a lot of information on the increasing amount of work being published in this field in Japan. The book has two sections: the first, 'Guide to books and audiovisual materials', gives a series of bibliographies arranged by subject or form of material; the second, 'Guide to original works of historical importance and their translation into other languages', is in alphabetical order of scientist. There is a combined author and subject index for the two sections. Some of the entries give either a list of the contents of the work or a critical annotation. It is a pity that there are not more of the latter; in particular, the section on biographies of physicists, where critical annotations would be of great value, is entirely without them.

The second guide, Thornton and Tully's *Scientific books libraries and collectors* (Library Association, 1971), is not just a bibliography and is not confined to physics. It aims to record the major books of every prominent scientific author before 1900, and the main editions of his work. Professional historians of science would doubtless quarrel with the inevitable selection that has taken place (for example, no mention of Ludwig Boltzmann); but this is undoubtedly a valuable introductory guide to the literature, and to bibliographies and libraries. It could equally be used just for reference: for example, finding out the original date and circumstances of publication of the works reprinted by Dover.

BIOGRAPHY

So far we have been concerned with the development of particular subjects. What about the people who made the developments? As was mentioned above, a list of biographies is given by Brush. But for very many physicists there does not exist a satisfactory biography. For others the work has been undertaken by a historian, who may give a reasonably adequate account of the life but lacks the necessary scientific background to do justice to the scientist's work. Even such a scholarly biography as Pearce Williams's *Michael Faraday* (Chapman and Hall, 1965) has been severely criticised on this score. Just as some of the most satisfactory work in history of science was done by working scientists such as Todhunter before the advent of the professional historian of science, so the best biographies tend to be those written by scientists. For example, Brewster's *Memoirs of the life, writings and discoveries of Sir Isaac Newton* (1855; Johnson, 1965) still provides one of the best accounts of Newton, except for certain aspects of his work in optics, where Brewster's physical critique has been invalidated by subsequent discoveries in physiological optics.

The gaps in our biographical knowledge have begun to be filled by the volumes of the *Dictionary of scientific biography* (Scribner, 1970–), edited by Gillispie. By the end of 1973 the first eight volumes, to the M's, had been published. The space devoted to a scientist is roughly proportional to his importance, and each entry concludes with a bibliography of the scientist's work and material about him. The standard is uniformly high. The other great dictionary from which brief biographical data on nineteenth and early twentieth century scientists can be found is the *Biographisch-literarisches Handwörterbuch zur Geschichte der exacten Wissenschaften* (J. A. Barth (varies), 1858–), started by Poggendorff. Besides giving biographical information, in the condensed form of a 'Who's Who' entry, the scientist's books and papers are listed. The dictionary has been published in several sequences of two or more volumes each, so that finding all the relevant material on one person may require looking in more than one sequence. Bd. 1 and 2 cover the period prior to 1858; Bd. 3 covers 1858–83; Bd. 4, 1883–1904; Bd. 5, 1904–22; Bd. 6, 1923–31; Bd. 7a and its supplement cover German scientists of all periods; and Bd. 7b, covering the period 1932–62 for non-German scientists, is in progress.

For scientists not yet included in the *Dictionary of scientific biography* details of books and articles about their work can be found in the *ISIS cumulative bibliography* (Mansell, 1971), edited by Magda Whitrow. This bibliography is based on the 'Critical bibliographies' established by George Sarton in his journal *ISIS* in 1913, and now produced as one issue of the journal every year. Each covers articles on all aspects of history of science from a very wide range of journals, and books, published in the preceding year. Despite their name, the bibliographies contain relatively few annotations, and these are of an indicative, not a critical, nature. The *ISIS cumulative bibliography* has been compiled from the critical bibliographies up to 1965, and the two volumes so far published cover: Vol. 1 Personalities A–J, and Vol. 2 Personalities K–Z and Institutions A–Z. Further volumes, giving a chronological and a subject approach, are planned.

PERIODICALS

The discussion of biography has taken us away from material which we might have in our own collections, and requires the larger resources of libraries. Similarly, if we are to pursue a particular subject in detail, we shall have to draw on libraries to give us access to periodical literature, both old science journals for papers which have not been reprinted in one of the collections, and journals in the history of science for recent studies. References to a scientist's papers can be found from a number of sources: Reuss' *Repertorium commentationum a societatibus litterariis* (1801–21; Franklin, 1962) covers articles appearing in the eighteenth century in the publications of scientific societies; Poggendorff has already been mentioned; a more complete coverage of the nineteenth century periodical literature is given by the *Royal Society of London catalogue of scientific papers, 1800–1900* (The Society, 1867–1924) and its successor, the *International catalogue of scientific papers* (Royal Society of London, 1902–21), covers the period 1901–14. More recent references can be traced by using the abstracting services described in Chapter 5. Before attempting to locate and obtain older scientific periodicals, it is worth checking whether the collected papers of the scientist whose work you are looking for have been published. A number of examples from the nineteenth century, such as the scientific papers of Maxwell and of Rayleigh, have been mentioned earlier in the chapter. It is now fairly common for a collected edition of a scientist's papers to appear a few years after his death. The collected papers of Bohr, Ehrenfest, Fermi, Lorentz, Minkowski, Pauli, Planck, Rutherford and von Neumann have all been published, and Readex have produced a microprint edition of Einstein's papers.

History of science journals have tended to remain general in their coverage of science, probably because the boundaries between different areas of science are never sharply defined and have varied with time; and the work of individual scientists often crosses any predetermined boundaries. It is only in the past few years, following increased interest in the history of science, that we have seen a growth in specialist societies and the development of a few specialist journals. Consequently, studies relevant to the history of physics are spread through a wide range of journals, and it is advisable to approach the periodical literature via indexes, primarily the *ISIS* critical bibliographies described above, or section 22 of *Bulletin signalétique: Histoire des sciences et des techniques* (CNRS, 1961–).

The Japanese were the first to produce a journal devoted to the history of physics: *Buturigakusi kenkyu* (Studies in the history of physics; Japanese Group for the History of Physics, 1958–). This is a mimeographed journal mainly in Japanese. The main access to Japanese work in the history of physics in Western languages is *Japanese studies in the history of science* (History of Science Society of Japan, 1962–). Truesdell's *Archive for history of exact sciences*

(Springer, 1960–   ) has been devoted primarily to physics, as have the annual volumes of *Historical studies in the physical sciences* (University of Pennsylvania Press, 1969–   ). *The natural philosopher* (Blaisdell, 1963–64) produced only three volumes. Information about history of physics projects, source materials, relevant conferences, and so on, is contained in the *Newsletter* of the American Institute of Physics Center for History of Physics (AIP, 1968–   ).

OTHER ASPECTS OF THE HISTORY OF PHYSICS

Two considerable areas associated with history of physics have not been covered in this survey: social and economic aspects of the history; and philosophy of science. It was felt that work in both areas is usually not concerned just with physics, and that, despite their interest, they should not be examined at the expense of subjects more directly relevant to modern physics. The most that can be achieved here is to whet the reader's appetite.

Those who think that debates on the place of 'pure' science are a modern phenomenon should consult Babbage's *Reflections on the decline of science in England* (1830; Irish UP, 1972). In it Babbage argued that there should be many scientists in the Government and that there should be Government grants to scientists. Babbage was one of those responsible for founding the British Association for the Advancement of Science, and an interesting picture of the Victorian attitude to the role of science, as well as of the scientific achievements of the period, can be found in the anthology of extracts from the presidential addresses to the Association edited by Basalla, *Victorian science* (Doubleday, 1970). One of the most illuminating modern accounts of the interaction of government and science is surely Greenberg's *The politics of American science* (Penguin, 1969).

Physics-based technologies are many and various. The mutual influence of theory and industrial technique in the development of thermodynamics is discussed in Cardwell's *From Watt to Clausius: the rise of thermodynamics in the early industrial era* (Heineman, 1971). One of the earliest electrical applications is described in Hubbard's *Cooke and Wheatstone and the invention of the telegraph* (Routledge, 1965). A good example of an account of the economic and social aspects of a physics-based technology is Margaret Gowing's official history of the United Kingdom Atomic Energy Authority, *Britain and atomic energy, 1939–1945* (Macmillan, 1964).

History and philosophy of science are mutually dependent. The former becomes a meaningless recital of facts unless there is implicit in it some understanding of the nature and purpose of scientific method. The latter remains sterile unless closely related to scientific practice in the form of case studies. An example of this interdependence is given by Mach's *The science of mechanics* (1883; Open Court, 1960), in which the history of the subject is subordinated to the attempt to analyse the underlying assumptions of his day. Duhem's *The aim and structure of physical theory* (1914; Atheneum, 1962) is a polemical attempt to make explicit the nature of physics, but contains more detailed studies of historical examples than many recent books on the philosophy of science. An exception is Kuhn's *The structure of scientific revolutions* (Univ. of Chicago Press, 1970).

BIBLIOGRAPHY

Primary sources

Ampère, André-Marie. *Théorie mathématique des phénomènes électrodynamiques, uniquement déduite de l'expérience.* Reprinted. Paris, Librairie Scientifique Albert Blanchard, 1958

Bacon, Francis. *Novum Organum.* Translated as: *The new organon.* Indianapolis, Liberal Arts Press, 1960

Bohr, Niels. *Collected scientific works.* New York, American Elsevier, 1971–

Boltzmann, Ludwig. *Populäre Schriften.* Reprinted. Würzburg, 'Journalfranz' Arnulf Liebing, 1972

Boltzmann; Ludwig. *Vorlesungen über Gastheorie.* Translated as *Lectures on gas theory.* Berkeley, University of California Press, 1964

Carnot, Sadi. *Réflexions sur la puissance motrice du feu ....* Translated as *Reflections on the motive power of fire.* New York, Dover Publications, 1960

Coulomb, Charles Augustin. *Mémoires de Coulomb.* Paris, Gauthier-Villars, 1884

de Broglie, Louis. *Recherches sur la théorie des quanta.* Reprinted. Paris, Masson, 1963

Descartes, René. *Principia philosophiae.* Translated as *Principes de la philosophie.* Paris, Librairie Philosophique J. Vrin, 1964

Drude, Paul. *Lehrbuch de Optik.* Translated as *The theory of optics.* New York, Dover Publications, 1959

Duhem, Pierre. *La théorie physique: son objet, sa structure.* Translated as *The aim and structure of physical theory.* New York, Atheneum, 1962

Eddington, Arthur Stanley. *Mathematical theory of relativity.* 2nd edition. CUP, 1924

Ehrenfest, Paul. *Collected scientific papers.* Amsterdam, North-Holland, 1959

Einstein, Albert. *Complete works.* London, Readex Microprint, n.d.

Einstein, Albert, and others. *Das Relativitätsprinzip.* Translated as *The principle of relativity.* New York, Dover Publications, 1952

Euler, Leonhard. *Leonhardi Euleri opera omnia.* Series secunda, Vols. 11/2, 12, 13. Zurich, Orell Füssli, 1960, 1954, 1956

Faraday, Michael. *Experimental researches in electricity.* Reprinted in 2 vols. New York, Dover Publications, 1965

Fermi, Enrico. *Collected papers*. 2 vols. Chicago, University of Chicago Press, 1962, 1965

Fourier, Joseph. *Théorie analytique de la chaleur*. Translated as *The analytical theory of heat*. New York, Dover Publications, 1955

Franklin, Benjamin. *Experiments and observations on electricity*. Reprinted. Cambridge, Mass., Harvard University Press, 1941

Galilei, Galileo. *Discorsi e dimonstrazioni matematiche intorno a due nuove scienze*. Translated as *Dialogues concerning two new sciences*. New York, Dover Publications, 1951

Gibbs, J. Willard. *Elementary principles in statistical mechanics*. Reprinted. New York, Dover Publications, 1960

Gibbs, J. Willard. *The scientific papers of J. Willard Gibbs*. 2 vols. New York, Dover Publications, 1961

Gilbert, William. *De magnete*. Translated. New York, Dover Publications, 1958

Heisenberg, Werner. *Die physikalischen Prinzipien der Quantentheorie*. Translated as *The physical principles of the quantum theory*. New York, Dover Publications, 1949

Hertz, Heinrich. *Untersuchungen über die Ausbreitung der elektrischen Kraft*. Translated as *Electric waves*. New York, Dover Publications, 1962

Huygens, Christiaan. *Traité de la lumière*. Translated as *Treatise on light*. New York, Dover Publications, 1962

Lagrange, Joseph Louis. *Mécanique analytique*. Reprinted in 2 vols. Paris, Librairie Scientifique et Technique Albert Blanchard, 1965

Landau, Lev D. *Collected papers*. Oxford, Pergamon Press, 1965

Lorentz, Hendrik Antoon. *Collected papers*. 9 vols. The Hague, Martinus Nijhoff, 1934–39

Lorentz, Hendrik Antoon. *Problems of modern physics*. Reprinted. New York, Dover Publications, 1967

Maxwell, James Clerk. *The scientific papers of James Clerk Maxwell*. New York, Dover Publications, 1965

Maxwell, James Clerk. *A treatise on electricity and magnetism*. Reprinted in 2 vols. New York, Dover Publications 1954

Minkowski, Hermann. *Gesammelte Abhandlungen*. 2 vols. Leipzig, Physiker Verlag, 1911

Nernst, Walther. *Die theoretischen und experimentellen Grundlagen des neuen Wärmesatzes*. Translated as *The new heat theorem*. New York, Dover Publications, 1969

Newton, Isaac. *Philosophiae naturalis principia mathematica*. Translated as *Mathematical principles of natural philosophy*. Berkeley, Calif., University of California Press, 1962

Newton, Isaac. *Opticks*. Reprinted. New York, Dover Publications, 1952

Pauli, Wolfgang. *Collected scientific papers*. 2 vols. New York, Interscience Publishers, 1964

Planck, Max. *Physikalische Abhandlungen und Vorträge*. 3 vols. Braunschweig, Friedr. Vieweg Sohn, 1958

Priestley, Joseph. *The history and present state of electricity, with original experiments*. Reprinted. New York, Johnson Reprint Corp., 1966

Rayleigh, John William Strutt. *The theory of sound*. Reprinted. New York, Dover Publications, 1956

Rayleigh, John William Strutt. *Scientific papers*. 3 vols. New York, Dover Publications, 1964

Rutherford, Ernest. *The collected papers of Lord Rutherford of Nelson*. 3 vols. London, Allen and Unwin, 1962–65

von Neumann, John. *Collected works*. 6 vols. Oxford, Pergamon Press, 1961–63

Collections and secondary sources

Babbage, Charles. *Reflections on the decline of science in England, and on some of its causes.* Reprinted. Shannon, Irish University Press, 1972

Basalla, George, *et al. Victorian science.* New York, Doubleday, 1970

Beyer, Robert T. *Foundations of nuclear physics.* New York, Dover Publications, 1949

Boorse, Henry A. and Motz, Lloyd. *The world of the atom.* New York, Basic Books, 1966

Brewster, David. *Memoirs of the life, writings and discoveries of Sir Isaac Newton.* 2 vols. Reprinted. New York, Johnson Reprint Co., 1965

Brush, Stephen G. *Kinetic theory.* 2 vols. Oxford, Pergamon Press, 1965, 1966

Brush, Stephen G. *Resources for the history of physics.* Hanover, New Hampshire, University Press of New England, 1972

Cardwell, Donald S. L. *From Watt to Clausius: the rise of thermodynamics in the early industrial age.* London, Heinemann, 1971

Ditchburn, Robert William. *Light.* 2nd edition. London, Blackie, 1963

Dugas, René. *La mécanique au XVIIe siècle.* Translated as *Mechanics in the seventeenth century.* Neuchatel, Éditions du Griffon, 1958

Dyson, Freeman J. *Symmetry groups in nuclear and particle physics.* New York, W. A. Benjamin, 1966

Ehrenfest, Paul and Ehrenfest, Tatiana. *Begriffliche Grundlagen der statistischen Auffassung in der Mechanik.* Translated as *The conceptual foundations of the statistical approach in mechanics.* Ithaca, N.Y., Cornell University Press, 1959

Gell-Mann, Murray and Ne'eman, Yuval. *The eightfold way.* New York, W. A. Benjamin, 1964

Gillispie, Charles Coulston. *Dictionary of scientific biography.* New York, Charles Scribner's Sons, 1970–

Gowing, Margaret. *Britain and atomic energy, 1939–1945.* London, Macmillan, 1964

Greenberg, Daniel S. *The politics of pure science.* New York, New American Library, 1967. Reprinted as *The politics of American science.* Harmondsworth, Penguin Books, 1969

Hesse, Mary B. *Forces and fields.* London, Nelson, 1961

Hubbard, Geoffrey. *Cooke and Wheatstone and the invention of the telegraph.* London, Routledge and Kegan Paul, 1965

Jammer, Max. *The conceptual development of quantum mechanics.* New York, McGraw-Hill, 1966

Jancel, Raymond. *Les fondements de la mécanique statistique classique et quantique.* Translated as *Foundations of classical and quantum statistical mechanics.* Oxford, Pergamon Press, 1969

Kabir, P. K. *The development of weak interaction theory.* New York, Gordon and Breach, 1963

Kilmister, Clive W. *Special theory of relativity.* Oxford, Pergamon Press, 1970

Kuhn, Thomas S. *The structure of scientific revolutions.* 2nd edition University of Chicago Press, 1970

Kuhn, Thomas S. and others. *Sources for history of quantum physics.* Philadelphia, American Philosophical Society, 1967

Ludwig, Gunther. *Wave mechanics.* Oxford, Pergamon Press, 1968

McGucken, William. *Nineteenth-century spectroscopy.* Baltimore, Johns Hopkins Press, 1969

Mach, Ernst. *Die Mechanik in ihrer Entwicklung historisch-kritisch dargestellt.*

Translated as *The science of mechanics*. LaSalle, Ill., Open Court Publishing Company, 1960

Mach, Ernst. *Die Prinzipien der physikalischen Optik*. Translated as *The principles of physical optics*. New York, Dover Publications, n.d.

Mendelssohn, Kurt. *The quest for absolute zero*. London, Weidenfeld and Nicolson, 1966

O'Rahilly, Alfred. *Electromagnetics*. Reprinted as *Electromagnetic theory*. 2 vols. New York, Dover Publications, 1965

Poggendorff, Johann Christian. *Biographisch-literarisches Handwörterbuch zur Geschichte der exacten Wissenschaften*. Leipzig, J. A. Barth (varies), 1858–

Reuss, Jeremias. *Repertorium commentationum a societatibus litterariis editarum secundum disciplinarum ordinem*. 16 vols. Reprinted. New York, B. Franklin, 1962

Rohrlich, F. *Classical charged particles*. Reading, Mass., Addison-Wesley, 1965

Ronchi, Vasco. *Storia della Luce*. Translated with additional material as *The nature of light*. London, Heinemann Educational Books, 1970

Sabra, A. I. *Theories of light from Descartes to Newton*. London, Oldbourne, 1967

Schwinger, Julian. *Selected papers on quantum electrodynamics*. New York, Dover Publications, 1958

Ter Haar, D. *The old quantum theory*. Oxford, Pergamon Press, 1967

Thornton, John L. and Tully, R. I. J. *Scientific books libraries and collectors*. 3rd edition. London, Library Association, 1971

Todhunter, Isaac. *A history of the mathematical theories of attraction and the figure of the earth*. Reprinted. New York, Dover Publications, 1962

Todhunter, Isaac and Pearson, Karl. *A history of the theory of elasticity and the strength of materials*. Reprinted in 3 vols. New York, Dover Publications, 1960

Tokaty, G. A. *A history and philosophy of fluid mechanics*. Henley-on-Thames, Foulis, 1971

Tomonaga, Sin-Itiro. *Quantum mechanics*. 2 vols. Amsterdam, North-Holland, 1962, 1966

Tricker, R. A. R. *The contributions of Faraday and Maxwell to electrical science*. Oxford, Pergamon Press, 1967

Tricker, R. A. R. *Early electrodynamics: the first law of circulation*. Oxford, Pergamon Press, 1965

Truesdell, Clifford Ambrose. *Six lectures on modern natural philosophy*. Berlin, Springer-Verlag, 1966

Van der Waerden, B. L. *Sources of quantum mechanics*. New York, Dover Publications, 1968

Westfall, Richard S. *Force in Newton's physics: the science of dynamics in the seventeenth century*. London, Macdonald; New York, American Elsevier, 1971

Whitrow, Magda. *ISIS cumulative bibliography: a bibliography of the history of science formed from ISIS Critical Bibliographies 1–90, 1913–1965*. 2 vols. London, Mansell, 1971

Whittaker, Edmund. *A history of the theories of aether and electricity*. 2 vols. London, Nelson, 1951, 1953

Williams, Leslie Pearce. *Michael Faraday*. London, Chapman and Hall, 1965

Yourgrau, Wolfgang and Mendelstam, Stanley. *Variational principles in dynamics and quantum theory*. 2nd edition. London, Pitman, 1960

8

Theoretical physics

Elliot Leader

'Theoretical Physics', as a single subject, no longer exists, except perhaps as an occasional course-title in the undergraduate curriculum. The extraordinary expansion of scientific activity since World War II has led to an ever-increasing degree of specialisation, so that today a physicist whose interest lies primarily on the theoretical side will find himself working in a much more narrowly defined area, such as Nuclear Theory, Solid State Theory, General Relativity, Statistical Mechanics, Elementary Particle Theory, etc. Thus, the contents of the present chapter must necessarily reflect the principal interests of the author, and will, therefore, deal mainly with Nuclear and Elementary Particle Theory,* although some attempt will be made to include more general aspects of Mathematical Physics and the most relevant parts of the literature on mathematical methods and tables.

Concomitant with the growth in scientific activity has been a veritable revolution in the methods of scientific communication. The tremendous rate of innovation and development has meant that the prepublication delay in traditional journals has become intolerable, and the cult of the *preprint* has emerged as a prime element in the field of communication. Originally, a preprint was a draft copy (perhaps even handwritten!) sent by an author to a few of his most respected colleagues, calling for their comments and criticisms, prior to submitting the article to a journal for publication. Today a preprint is the final version of an author's work, beautifully typed,

---

* *Editorial note.* To partially make up for these gaps a supplementary list of works has been added at the end of the chapter. That brief bibliography is clearly not addressed to specialists in each field; hopefully it will serve as a first guide for sudents. Of course the general treatises described in Chapter 4 include these areas, often in a very authoritative manner.

lavishly illustrated, often bound, and sent out in hundreds of copies by the author's institution at the same time as it is submitted for publication. This method of communication is, by now, so firmly established and so widespread throughout the world that not only can it not be ignored as simply a 'private' activity, but it must even be accorded a leading position as one of the principal modes of present-day communication.

Another consequence of the demand for rapid publication has been the appearance of *letter journals*. Originally, most journals contained a 'letters to the editor' section for short communications, for criticisms of published articles and, often, for lively exchanges of opposing points of view. Under the pressure for speedier publication, several journals have given birth to separate, frequently issued off-spring, containing only letter-length articles. It was the hope and intention that authors would restrict themselves to 'urgent' com-munications, to matters which they felt deserved immediate and widespread dissemination among the physics community. With such an utterly subjective criterion it is not surprising that the letter jour-nals have to some extent become the carriers of short, and sometimes unintelligible, summaries of an author's work, or even, on occasion, of what amounts to an announcement of discovery with little explana-tion. Often, lengthier follow-up articles in the regular journals are promised, but once the excitement of the initial proclamation has passed, lethargy sets in, with the result that long periods elapse before the complete write-up appears; or it may never materialise at all! Despite this abuse, it must be acknowledged that the letter journals serve a very useful purpose in communicating newly found experimental results. The absence of sufficient information does mean that on occasion one may misinterpret the significance or reliability of a particular result, but experience soon teaches one how to exercise caution.

The explosion of scientific activity has also caused the emergence of a large number of new journals; to such an extent, in fact, that no single physicist can hope even to scan the complete literature adequately. To compensate for this, one may consult the *abstracting journals*, which list title, author, and a brief summary of recently published material. However, the abstracts appear some months after the article appears in a journal and, therefore, many months after the appearance of the preprint.

A new twist is provided by the *current awareness* or *alerting* journals, which either publish *advance* abstracts of articles due to appear later in the regular journals or simply publish what is basically a list of the titles and authors of articles currently appearing in the regular journals. Since these have appeared on the scene only recently, it is too early to assess whether they will assume a really important role.

One further consequence of the mushrooming of scientific activity and, more particularly, of the rate of development of new ideas and theories is the difficulty of producing timely and relevant textbooks. The traditional textbook, as the culmination of mature and considered reflection over a long period, is simply out of date by the time it appears. To some extent the role of the textbook, as providing a detailed pedagogical development of a given field, has been taken over (1) by various series of paper-backed monographs, often constituted from a set of recently written lecture notes, and rapidly put together and produced, or (2) by the published proceedings of the numerous summer schools and institutes that are now held annually in many different countries, and at which lectures on and discussions of research level topics take place. The level of the lectures is set by the fact that the schools aim usually to cater for the equivalent of a British graduate student in his second year of research. These summer school proceedings are often the best and most up to date sources, in depth, on current research topics.

In general terms, theoretical physicists tend to make use of relatively recently published material and have far less need than, say, chemists to search out very old material. As a result, they are not very adept at searching the literature, and it often happens, as a consequence, that a new result turns out to be the rediscovery of something long buried in the literature. There is no easy solution to this problem, but every graduate student should make a determined effort to get himself into the habit of regularly and conscientiously perusing the literature. It is only the familiarity that comes from *active* participation that can lead to a confident and efficient handling of the wealth of research material hidden in the journals.

In the following we shall list the most important literature sources in current use. They are classified according to type, and brief comments are added where necessary.

PREPRINTS

All the major universities and research laboratories now send out preprints on behalf of their research workers. To obtain regular information of their research output, it is necessary to establish oneself on their mailing lists. The huge scale of the operation now almost precludes the possibility for an *individual* to be a regular recipient. Generally, departments or research groups have banded together to set up 'preprint libraries' or 'preprint collections', with a designated secretary (often a research student) who handles, sorts and catalogues the incoming material, and ensures that it is adequently displayed. In this way the material is mailed to the group as a whole, via the preprint secretary. Often an institute will request a reciprocal arrangement from a group, seeking to be placed on its mailing list. Often, too, a large institute will classify its mailing list according to broad areas of subject interest.

Generally speaking, it seems that in universities it is more efficient and effective to organise a preprint library *inside* a department and divorced from the main university library, but, on the contrary, at research laboratories preprint collections within the main library seem to function excellently.

For a newly constituted research group, or for a small university, it is no simple matter to ensure that it has access to all the relevant material, appearing, perhaps, from hundreds of different sources. Very helpful in this problem are the *lists of preprints received* sent out at regular intervals, usually every week, by some of the major laboratories. The titles and institutes are:

Preprints in Particles and Fields (issued by the Stanford Linear Accelerator Center, Library, P.O. Box 4349, Stanford, California 94305, USA).

Preprint Registry for High-Energy and Elementary Particle Physics (issued by the International Centre for Theoretical Physics, Miramare, Trieste, Italy)

Preprints and Reports (issued by the European Centre for Nuclear Research (CERN), Library, 1211 Geneva 23, Switzerland)

Recent Additions to the Rutherford Laboratory Library (issued by the Rutherford High Energy Laboratory, Library, Chilton, Didcot, Berks., United Kingdom)

It is safe to say that, between them, these internationally known institutions receive preprints from every reputable source in the world. The Stanford list also provides a unique service by means of its *Anti-preprint List*, which gives the journal reference for previously listed preprints as and when they are published in the regular journals. The above lists can be obtained on a regular basis upon request to the institutions concerned.

THE LETTER JOURNALS

The two major letter journals most relevant for theoretical papers on topics close to Nuclear or Elementary Particle Physics are: *Physical Review Letters* (published weekly by the American Physical Society) and *Physics Letters B* (published fortnightly by North-Holland).

The coverage of topics is very broad in *Physical Review Letters*; the *B* section of *Physics Letters* deals exclusively with papers on Nuclear and Elementary Particle Physics. It also has a larger proportion of papers on theoretical topics. Both journals are especially useful for learning about new experimental results. However, the results are very often presented in small-scale graphical form, and detailed numerical results, if urgently required, must be requested directly from the authors.

Two other letter journals of some interest are: *Lettere al Nuovo Cimento* (published by the Società Italiana di Fisica), which deals almost exclusively with short theoretical notes, and *JETP Letters* (a translation by the American Institute of Physics, of the Russian *JETP Pis'ma v Redaktsiyu*), which is very similar in style to *Physical Review Letters*, but which handles, in practice, papers originating exclusively in the USSR and its satellite countries.

Although it is possible to ignore the letter journals completely, if one is working on a very mathematical or fundamental topic in which the time-scale for interesting developments is years, rather than weeks, it nevertheless remains a fact that one will thereby be cutting oneself off from an important source of the excitement and the spirit of discovery that should pervade all research activity in physics.

THE PRINCIPAL JOURNALS

As mentioned in the introduction, there are far too many journals for any single physicist to hope to cover. A few of the journals do tend to play a somewhat more prominent role than others, but from time to time there will appear an article of immense importance in a relatively little-read journal. The only answer to this problem is to make regular use also of the *abstracting journals* (to be discussed below).

The main journals to which an elementary particle or nuclear physicist is likely to find himself turning are, in alphabetical order:

Annals of Physics (fortnightly, Academic Press), which covers all areas of basic physics research. It tends to publish long, and often definitive, articles on new topics.

Nuclear Physics B (fortnightly, North-Holland) is devoted entirely to Elementary Particle Physics. In recent years it has become very much *the* European journal in the field. The emphasis is somewhat towards experimental and phenomenological papers.

Nuovo Cimento A (twice monthly, Società Italiana di Fisica) carries articles on Quantum Theory, Elementary Particle Physics and Nuclear Physics. Also, occasionally, it publishes articles on the more fundamental problems in theoretical physics.

Physical Review D (fortnightly, American Institute of Physics) probably enjoys the highest reputation for containing significant and important contributions on many topics; a reputation which in some measure is due to its very strict system of refereeing of submitted papers. The subject matter is limited to Particles and Fields, and includes, under the latter, also General Relativity.

Progress of Theoretical Physics (monthly, in English, Research Institute for Fundamental Physics and the Physical Society of Japan) has a very broad coverage of all areas of theoretical physics. Contributions are mainly, but not exclusively, from Japanese institutes.

Soviet Journal of Nuclear Physics (translation by the American Institute of Physics of the Russian monthly *Yadernaya Fizika*) covers both Nuclear and Elementary Particle Physics and is somewhat analogous to *Physical Review*. However, many articles tend to be too brief to be easily comprehensible.

Soviet Physics Doklady (a translation by the American Institute of Physics of the Physics Section of the Proceedings of the Academy of Sciences of the USSR) has mainly classical theoretical physics, but occasional articles on Elementary Particles.

There are several journals that are devoted to a more general coverage of physics as a whole, some of which occasionally contain articles on nuclear and elementary particle physics. A selection follows:

Acta Physica Austriaca (monthly, the Österreichischen Akademie der Wissenschaften) has articles in German and in English. The subject matter is fairly broad.

Acta Physica Polonica B (monthly, Polska Akademia Nauk) often has articles in English on Relativity and Field Theory, as well as on Nuclear and Elementary Particle Theory.

Journal of Physics A (monthly, The Institute of Physics) deals with Mathematical, Nuclear and more general Theoretical Physics.

Physical Review A (monthly, American Institute of Physics) covers the field of General Physics.

Soviet Physics JETP (a translation by the American Institute of Physics of the Russian monthly *Zhurnal Eksperimental'noi i Teoreticheskoi Fiziki*) principally covers topics in Atomic Physics and in General Physics, including Plasma and Solid State Physics.

In all the above journals the emphasis is on the physics and the physical ideas. Mathematics usually plays a secondary role; it is simply a part of the language of physics, a necessary and vital tool. There are some important journals where, on the contrary, the emphasis is on new mathematical ideas and techniques. Generally, there is stress upon mathematical rigour, and it can easily happen that the concepts and notation are outside the repertoire of the average physicist. But quite often it happens that some important technique gradually gains broader recognition and eventually becomes a part of the standard physicist's armoury.

It goes without saying that breadth and depth of mathematical knowledge are a magnificent asset to a theoretical physicist, so that a regular perusal of the following journals will certainly be rewarding:

Annales de l'Institut Henri Poincare A (8 issues p.a., Gauthier-Villars) has articles of a very mathematical nature and requiring a high degree of mathematical sophistication. The main topic of interest to Elementary Particle Theory is Axiomatic Field Theory.

Communications in Mathematical Physics (quarterly, Springer) is only a little less demanding in mathematical background than the previous journal, and concentrates on Quantum Mechanics, Quantum Field Theory, Statistical Mechanics and General Relativity.

Journal of Mathematical Physics (monthly, American Institute of Physics) is perhaps the most popular journal dealing with mathematical techniques. It does not usually require a very sophisticated knowledge of mathematics. The material is of a very broad nature and is relevant to a very wide field of physics.

In recent years the computer has come to play an enormously important role in physics, not just as an arithmetical aid but also as a vital structural component in the controlling of experimental apparatus, in the handling and organising of data, in the automatic analysis of bubble-chamber film and spark-chamber output and, more recently, in the manipulation and simplification of complicated algebraic expressions. An exciting development is the application of these methods to the evaluation of Feynman diagrams in Field Theory. Program writing and software development have reached a high degree of sophistication, and specialists in this activity are catered for in the *Journal of Computational Physics* (monthly, Academic Press).

The acceptance of Quantum Mechanics and, more generally, of quantum theoretic ideas, into the framework of theoretical physics, some 50 years ago, brought with it a host of new logical and philosophical problems. Although the *vast majority* of physicists use quantum mechanical ideas quite routinely, very few indeed concern themselves with the unsolved epistomological difficulties. However, there are a few groups working in this borderline region between physics and philosophy, and a forum for discussion of their activity is available in the journal *Foundations of Physics* (quarterly, Plenum Press).

THE ABSTRACTING JOURNALS

There are two major abstracting journals of relevance. Both appear fortnightly and cover roughly the same journals. The coverage is excellent and a huge number of journals is scanned. The abstracts are classified according to subject matter and the entry contains the title, author, his institution and its address, the reference to the journal in which the article is published and the abstract itself, i.e. a short résumé of the contents of the paper and of the main results derived. The abstracts can be traced via a subject index and an author index. In both journals the published abstract is given a serial reference number, and all indexes use this for reference. A cumulative index is published twice yearly. The journals are:

Nuclear Science Abstracts (United States Atomic Energy Commission), which deals principally with papers of direct interest to Nuclear Physics and Elementary Particle Physics (see p. 218).

Physics Abstracts (Institution of Electrical Engineers). This is the continuation of the long established *Science Abstracts Series A*. It covers a very broad area of physics.

In addition to the subject and author indexes, *Nuclear Science Abstracts* provides indexing according to 'corporate author', i.e. the author's institution or laboratory, and according to certain report series. There is also coverage of United States Government Laboratory reports.

The 6-monthly cumulative index of *Physics Abstracts* has several special headings: Bibliography Index, which refers to articles with a significant list of references or to specifically bibliographic papers; Book Index; Patent and Report Index; Conference Index, which lists conferences for which published proceedings have appeared, and which also gives reference to the individual papers in the proceedings.

A derivative, shortened version of *Physics Abstracts* is *Current Papers in Physics* (Institution of Electrical Engineers), which lists the title, author, etc., but not the *abstract* itself, for all the entries in *Physics Abstracts*. This is a much less expensive journal than *Physics Abstracts*, but its value seems somewhat marginal.

The last word in abstracting is provided by the new 'current aware-ness' journals which seek to alert the reader to the existence of an article well in advance of its actual publication in a regular journal. *Current Physics Advance Abstracts* (AIP) was an attempt to do this, by publishing author abstracts several months before publication of the articles based on material accepted for publication in AIP journals. However CPAA did not last long and has been replaced by *Current Physics Index* (see p. 52).

Current Contents (weekly, Institute for Scientific Information), simply presents reprints of the Tables of Contents of many currently appearing journals in the whole field of Physical and Chemical Sciences. An author index and address directory are also provided, to enable one to write directly to the author for preprints or reprints.

The role and importance of these newer journals is very much dependent on the individual's situation. If one has access to a well-stocked, efficiently supplied library, then there is no great advantage to using the latter journals, since one can just as well scan the contents of the regular journals themselves; although it is marginally more con-venient to have all the contents collated into one slim volume. How-ever, in the present climate of financial stringency many libraries have been forced to cease taking the less heavily used journals. In this situation the 'Current' journals, which are inexpensive, serve as a useful starting point. A desired journal can then be obtained via the Inter-Library Loan service.

REVIEW JOURNALS

Several journals are devoted to publishing long survey articles that aim to give a comprehensive and up-to-date picture of a whole re-search topic. A good review article is of inestimable value. It should give an authoritative and well-reasoned judgement on the reliability and significance of experimental data, and should put into perspective the various aspects of a theory, bringing out clearly the meaning and relationship of the ideas involved.

Although most of the articles contain a lengthy introduction, they are not usually suitable for pedagogic use and their level is fairly sophisticated. However, a conscientious reviewer will have provided sufficiently complete references to background material, so that an advanced graduate student can use his article as a starting point for a study of the subject. Most of the journals try to keep their coverage of topics quite broad, but some, especially those appearing annually, are rather specialised. The most useful for Nuclear and Elementary Particle Physics are:

Annual Review of Nuclear Science (Annual Reviews), which covers everything from low-energy nuclear phenomena to high-energy elementary particle reactions. A useful reprint ordering service is offered, and offprints of individual articles can be purchased. A cumulative index of contributing authors and chapter titles is provided. For access to articles published from 1963 to 1972, reference should be made to Volume 22 (1972). For the period prior to 1963 refer to Volume 12 (1962). A few of the 'classic' articles on Elementary Particle Physics are listed below.

1958: 'Invariance principles of nuclear physics', by G. C. Wick

1960: 'Nucleon–nucleon scattering experiments and their phenomenological analysis', by M. H. MacGregor, M. J. Moravcsik and H. P. Stapp

1965: 'Spin and parity determination of elementary particles', by R. D. Tripp

1965 and 1966: 'Weak interactions', by T. D. Lee and C. S. Wu (corrections appear in the 1967 volume)

1969: 'Boson resonances', by I. Butterworth

1969: 'Pion–nucleon interactions', by R. G. Moorhouse

1970: 'The present status of quantum electrodynamics', by S. J. Brodsky and S. D. Drell

1972: 'Deep inelastic electron scattering', by J. I. Friedman and H. W. Kendall

1972: 'The nucleon–nucleon effective range expansion parameters', by H. P. Noyes

Physics Reports (approximately monthly, North-Holland) is a relatively new journal, but has very quickly established itself as a source of comprehensive, in-depth studies on up to the moment topics. The coverage is broad. Some examples, relevant to Elementary Particle Physics, are:

Volume 1 c (1971)

No. 4: 'Regge theory and particle physics' by P. D. B. Collins. Contains all the major ideas in this active field.

No. 7: 'Particle production in high energy hadron collisions', by L. Van Hove. Surveys empirocal facts and theoretical ideas on multiparticle production reactions

Volume 2 c (1972)

No. 1: 'Theoretical models of diffraction scattering', by F. Zachariasen

No. 3: 'Meson–meson scattering', by J. L. Petersen

Volume 3 c (1972)

No. 1: 'The possible role of elementary particle physics in cosmology', by R. Omnes

No. 3: 'The proton–neutron mass difference problem and related topics.' Surveys many different approaches to this fundamental, unsolved problem.

No. 4: 'Recent developments in the comparison between theory and experiments in quantum electrodynamics', by B. E. Lautrup, A. Peterman and E. de Rafael. A very clear and physically motivated review.

Volume 4c (1972)

No. 5: 'Leptonic decays of hadrons', by L. M. Chounet, J. M. Gaillard and M. K. Gaillard

Volume 5 c (1972)

No. 3: 'High energy theorems for strong interactions and their comparison with experimental data', by S. M. Roy

No. 5: 'Multiple production of hadrons at cosmic ray energies (experimental results and theoretical concepts)', by E. L. Feinberg

To appear in 1973:

'Renormalisation group', by K. G. Wilson

Volume 8 c, No. 2: 'The parton picture of elementary particles', by J. Kogut and L. Susskind

Reviews of Modern Physics (quarterly, American Institute of Physics) has a broad spectrum of interest, from the very abstract to the applied. It aims to provide detailed reviews, and to evaluate significant topics in physics. Its quality fluctuates a great deal, but it does, on occasion, produce articles of great merit. A helpful list of titles of reviews appearing in other journals is appended. Some useful articles are:

1963: 'The octet model and its Clebsch–Gordan coefficients', by J. J. de Swart

1964: 'Tables of Clebsch–Gordan coefficients for SU(3)', by F. Chilton and P. McNamee

1965: 'The conceptual basis and use of the geometric invariance principles', by M. F. Houtappel, H. Van Dam and E. P. Wigner

1966: 'Unitary groups: representations and decompositions', by C. Itzykson and M. Nauenberg

1968: 'Phase and angle variables in quantum mechanics', by P. Carruthers and M. M. Nieto

1969: 'Effective Lagrangians and field algebras with chiral symmetry', by S. Gasiorowicz and D. A. Geffen

1970: 'Crossing matrices for SU(2) and SU(3)', by C. Rebbi and R. Slansky

1970: 'Models for high energy processes', by J. D. Jackson

1972: 'High energy multiparticle reactions', by W. R. Frazer *et al.* A very clear, not too technical review

Also of great importance is the 'Review of Particle Properties', which appears as a regular annual feature in *Reviews of Modern Physics*. This is an authoritative survey and evaluation of the world data on the properties of elementary particles and on the discovery of new particles. It is written by an international team of physicists and is now recognised as the world standard in its field. In recent years the Review has also been published in *Physics Letters B*.

There are many other review journals which from time to time carry first-rate articles of direct relevance to the nuclear and elementary particle field. It is without question worth while to keep one's eye on their Tables of Contents.

Essays in Physics (annually, Academic Press) tries to offer general surveys with emphasis on unifying ideas which link several areas of physics. This is not essential reading, but it can be very stimulating. *Fortschritte der Physik* (monthly, Akademie, for the Physikalische Gesellschaft in der Deutschen Demokratischen Republik) contains articles in German and English.

Progress in Elementary Particle and Cosmic Ray Physics (North-Holland) has long reviews that may, in some cases, be definitive, but in others are simply progress reports.

Progress in Nuclear Physics (annually, Pergamon Press) concentrates on recent developments in the nuclear field and on guiding the reader through the primary literature. It contains a cumulative reference index to earlier volumes.

Reports on Progress in Physics (monthly, Institute of Physics) carries non-specialist reviews, generally with a long historical introduction. This is a very useful journal for keeping oneself abreast on subjects outside one's own research interest.

Rivista del Nuovo Cimento (published by the Società Italiana di Fisica) contains rather short review-type articles, usually on a very narrowly defined topic.

Soviet Physics Uspekhi (a translation by the American Institute of Physics of the Russian *Uspekhi Fizicheskikh Nauk*) is the Russian analogue of *Reviews of Modern Physics*, but its articles tend to be somewhat shorter.

Supplement of Progress of Theoretical Physics (published in English by the Research Institute for Fundamental Physics and the Physical Society of Japan) surveys developments in many branches of theoretical physics, but the principal emphasis is on work produced by Japanese scientists, or by visitors to Japan.

LECTURE NOTE SERIES AND PROCEEDINGS OF 'SCHOOLS'

These volumes are probably the best source of advanced *pedagogical* material. They usually contain long, gradually developed expositions of topics of current interest, prepared primarily as lecture material for first- and second-year research students. In some cases the published version has benefited from the feedback from the student audiences. The various Summer Schools, Summer Institutes, Winter Schools, etc., generally gather annually and devote themselves to a different subject each year. There are not usually cumulative indexes, and to make matters more difficult, some of the Schools seem to choose a different publishing firm almost every year. However, the value of the articles cannot be overstated, and every research student should spend some time examining these volumes. The major publications are listed below. The dates given refer to the period when the school was held.

Brandeis University Summer Institute in Theoretical Physics (Gordon and Breach, New York, or MIT press). Particularly interesting volumes or articles are:

1965: Vol. 1, 'Axiomatic field theory'
Vol. 2, 'Particle symmetries'
1967: Vol. 1, 'High energy behavior of scattering amplitudes', by F. E. Low, which has a good treatment of fixed poles in the complex J-plane.
1970: Vol. 1, 'Perturbation theory anomalies', by S. L. Adler
'Dynamical applications of the Veneziano formula', by S. L. Mandelstam
'Dynamics and algebraic symmetries', by S. Weinberg
'Local operator products and renormalization in quantum field theory', by W. Zimmerman

Cargèse Lectures in Theoretical Physics (Gordon and Breach)—the proceedings of the Cargèse summer school. Some articles of interest are:

1964: 'Electron scattering', by L. I. Schiff, which covers scattering on nucleons and nuclei. The calculations are done very explicitly.

1966: 'The Wigner–Weisskopf method', by J. Bernstein. Applies the ideas to several branches of physics.

1967: Several pedagogical articles with a modern approach to problems in quantum electrodynamics, especially

'Renormalization theory', by D. R. Yennie

'Summation methods for radiative correction', by K. E. Eriksson

'Some applications of coherent states', by T. W. B. Kibble

1970: Articles on both the mathematical properties and physical applications of 'Padé approximants'.

Lectures in Theoretical Physics (Gordon and Breach), which are the Proceedings of the Theoretical Physics Institute of the University of Colorado at Boulder. Articles or volumes of special interest are:

1958 (Vol. I): 'High energy collision theory', by R. J. Glauber, which develops the now famous Glauber Theory in a very clear fashion.

1960 (Vol. III): 'Quantum theory of collision processes', by R. Haag, in which a formal and rigorous approach is developed.

1964 (Vol. VII A): 'Lorentz group'. Many articles on various aspects of the Lorentz Group, its structure, representations, role in particle physics, etc.

1966 (Vol. IX A): 'Mathematical methods of physics': several papers on the more mathematical aspects of Symmetry Groups, Lie Groups, Non-Compact Lie Algebra, etc.

1968 (Vol. XI D): 'Infinite momentum frames and particle dynamics', by L. Susskind, which introduces ideas involved in studying dynamical development of a system in an 'infinitely' fast moving reference frame.

1969 (Vol. XII B): 'Daughters, conspiracies, Toller poles: some problems in the Reggeization of relativistic processes' by E. Leader, which gives a pedagogical introduction and survey of different approaches to the problem.

1970 (Vol. XIII): 'Introduction to the De Sitter and conformal groups and their physical application', by A. O. Barut. Also, several papers on the more technical aspects of these groups are included.

Proceedings of the Canadian Summer Institute in Nuclear and Particle Physics (McGill University, Montreal). Some articles of interest are:

1973: 'Two component models of particle production', by C. Quigg

1973: 'Lepton–hadron interactions', by B. Barish, discusses, in particular, the newest developments in neutrino–hadron interactions and the possible evidence for neutral currents.

Proceedings of the International School of Physics 'Enrico Fermi' (Academic Press). One of the oldest and best-established schools, the topics range over many areas in physics. Interesting articles are:

1963, Course XXIX: 'The proof of dispersion relations', by M. Froissart, which gives a clear outline of the main steps used in the standard (and very complicated!) proof of dispersion relations.

1964, Course XXXIII: 'Strong interactions'. Excellent articles by R. H. Dalitz, D. H. Miller and R. D. Tripp on the analysis of resonance production and the determination of the spin and parity.

1967, Course XLI: 'Current algebra at small distances', by J. Bjorken. This has a nice physical description of ideas which have recently become of great interest in inelastic electron scattering at high-momentum transfers and in questions of the 'constituents' of nucleons.

1969, Course XLVI: 'Theory of e^+–e^- annihilation at high energy'— a very thorough and detailed discussion.

1970, Course XLVII: 'Local field theory'. Several articles on the axiomatic approach to quantum field theory.

Proceedings of the International School of Physics 'Ettore Majorana' (Academic Press). Articles of interest are:

1964: 'Vector and axial currents under first order symmetry breaking', by R. Gatto—a pedagogical discussion of Gatto's theorem on non-renormalisation.

1965: 'Difficulties of relativistic U(6)', by J. S. Bell—a very clear treatment of a subtle subject.

1966: 'An introduction to unitary symmetry', by S. Coleman—an individualistic and informative approach.

1968 (Part A): 'New developments in current algebra', by S. Fubini, which has a lucid discussion of disconnected graphs.

1968 (Part A): 'Lectures on the quark model', by G. Morpurgo. This is interesting and wide-ranging, but not simple.

1969 (Part A): 'Inelastic electron scattering, asymptotic behaviour and sum rules', by S. D. Drell—very good as a pedagogic intro-duction.

1969 (Part A): 'Multiperipheral dynamics', by M. L. Goldberger

1969 (Part A): 'Acausality', by S. Coleman. This is a rambling, but extremely stimulating, look at some unusual questions.

1970 (Part A): 'Rigorous results from field theory and unitarity', by A. Martin

1970 (Part A): 'Narrow resonance models compatible with duality and their development', by G. Veneziano—a very intelligible discussion of an incredibly complicated subject.

1970 (Part B): 'Review of experimental results on the $\pi\pi$ and Kπ interactions', by P. E. Schlein

Proceedings of the Internationale Universitätswochen für Kernphysik at Schladming (published as a supplement to *Acta Physica Austriaca* by Springer). Some useful articles follow.

1968: 'Discrete symmetries in elementary particle physics', by J. Nilsson, gives a detailed and clear introduction to the discrete symmetries and illustrates their use.

1969: 'Finite energy sum rules—use and interpretation', by D. Horn —a simple introduction by one of the inventors of the idea.

1969: 'Miscellaneous topics related to the annihilation of an electron–positron pair into mesons', by M. Gourdin. Full of useful formulae on various types of annihilation, this article also looks at the vector-meson dominance model and at ω–ϕ mixing.

1970: 'Introduction to the use of non-linear mapping techniques in S-matrix theory', by D. Atkinson

1970: 'Aspects of chiral symmetry', by B. Renner

1972: 'Survey of high energy inelastic models and kinematical constraints for inclusive processes', by E. Predazzi

Proceedings of the Les Houches Summer School (Gordon and Breach). Of particular interest is the following volume:

1965: 'High energy physics'. Articles by G. C. Wick on Invariance Principles, R. H. Dalitz on the Quark Model, etc.

Proceedings of the Rencontre de Moriond (Société Polygraphique Mang, Paris). One of the few gatherings specifically aimed at bringing together theorists and experimentalists, its proceedings carry medium-length, research-level talks on topics of current interest.

Proceedings of the Scottish Universities Summer School in Physics (Academic Press). A volume of particular interest is:

1970: 'Hadronic interactions of electrons and photons', which contains several excellent articles on photoproduction, electroproduction, SU(3) properties of the electromagnetic current, electromagnetic mass differences, vector dominance, high-energy photon and neutrino interactions with nuclei.

Springer Tracts in Modern Physics (Springer), the continuation of *Ergebnisse der Exacten Naturwissenschaften*
These tracts sometimes cover the proceedings of German schools,

e.g. the International Summer School for Theoretical Physics, Karlsruhe. Interesting volumes are:

Vol. 57 (1971): Several articles on 'Strong interactions' dealing with analyticity and crossing, and with dual models for N-point functions.

Vol. 59 (1971): 'Symposium on meson-photo and electroproduction at low and intermediate energies', which has a comprehensive discussion of both theoretical and experimental techniques.

CONFERENCE REPORTS

There are several conferences held with some regularity where new results are presented and the very latest developments are discussed.

Often the venue for the conference changes from year to year, and may be in a different country each time, so the publication of the proceedings are handled locally. The volumes of a conference series are generally shelved together in a library. Their publication is announced in the abstracting journals.

The contents, typically, are at an advanced research level, and are unsuitable as an introduction to a subject. They are invaluable, however, for short-term, up-to-date surveys of the state of activity in any given research area.

Some of the best-established series are listed below.

International Conference on High Energy Physics (held every 2 years)

International Conference on Elementary Particle Physics (held every 2 years; alternates with the High Energy Conference)

International Conference on High Energy Collisions ('Stonybrook Series', held annually)

TEXTBOOKS AND MONOGRAPHS

It is quite impossible to attempt a systematic survey of the hundreds of volumes whose contents would be of relevance to a theoretical physicist, even if one restricted oneself to Nuclear or Elementary Particle Physics. What follows, therefore, is a list of personal favourites, of texts which have, in the author's opinion, some special characteristics of clarity, or of depth.

The most outstanding modern coherent 'course' on theoretical physics is found in the nine-volume treatise of Landau and Lifshitz, translated from the Russian (see p. 39 for full details).

This is a magnificent work, written with great clarity, and emphasising the unity of the physical ideas that interweave the whole set. The course can be started at the advanced undergraduate level, and the attainment of a detailed knowledge of the contents would provide one with a superb general knowledge of the whole field of theoretical physics.

At a more specialised level, the following have proved exceptionally useful:

Merzbacher, E. *Quantum mechanics*. Wiley, 1961. Has a very good balance between the development of the theory and its applications. It really teaches one how to *use* quantum mechanics.

Dirac, P. A. M. *The principles of quantum mechanics*. Clarendon, 4th edition. The classic, deep and beautifully lucid exposition of the fundamental underlying structure of quantum mechanics.

Bethe, H. A. and Jackiw, R. *Intermediate quantum mechanics*. Benjamin, 1968. Advanced aspects of mainly non-relativistic quantum mechanics.

De Alfaro, V. and Regge, T. *Potential scattering*. North-Holland, 1965. An excellent, high-level treatment of the more mathematical aspects of non-relativistic potential scattering.

Calogero, F. *Variable phase approach to potential scattering*. Academic Press, 1967. Describes an almost unknown but interesting and useful method of solving non-relativistic scattering problem.

Hagedorn, R. *Relativistic kinematics*. Benjamin, 1963.

This is a beautifully clear, very simple and instructive book.

Bjorken, J. D. and Drell, S. D. *Relativistic quantum mechanics* and *Relativistic quantum fields*. McGraw-Hill, 1964. The two volumes share a common approach based on the use of propagators. The treatment of renormalisation is very thorough. This is probably the best modern treatment of quantum field theory.

Henley, E. M. and Thirring, W. *Elementary quantum field theory*. McGraw-Hill, 1962. Very good exposition of the physical ideas, which so often are hidden in the mathematical apparatus of field theory.

Gasiorowicz, S. *Elementary particle theory*. Wiley, 1967.

Martin, A. D. and Spearman, T. D. *Elementary particle theory*. North-Holland, 1970. Both these volumes give a modern, and broad introduction to theoretical techniques used in Elementary Particle Physics. Martin and Spearman are very explicit in the steps and details of their calculations.

Roman, P. *Theory of elementary particles*. North-Holland, 1960. Although somewhat out of date, this has a good pedagogical treatment of field equations, and selection rules from invariance principles and also a very explicit, introductory discussion on orthogonal groups.

Perkins, D. H. *Introduction to high energy physics.* Addison-Wesley, 1972. A good, general introduction, intelligible also to advanced undergraduates.

Pilkuhn, H. *The interactions of hadrons.* North-Holland, 1967. A modern introduction to the whole field of strongly interacting particles. The approach is largely phenomenonological, and avoids all detailed mathematical arguments.

Eden, R. J. *High energy collisions.* CUP, 1967. A useful, introductory text, covering an amalgam of phenomenology and theory, the latter being based mainly on analytic properties of scattering amplitudes.

Lichtenberg, D. B. *Unitary symmetry and elementary particles.* Academic Press, 1970. An excellent pedagogical text.

At a very specialised level, and requiring a good knowledge of relativistic quantum mechanics, there are several monographs on topics of current research interest.

Eden, R. J., Landshoff, P. V., Olive, D. I. and Polkinghorne, J. C. *The analytic S-matrix.* CUP, 1966. This is the most detailed treatment of the analytic properties of scattering amplitudes as deduced from field theory, and of the high-energy behaviour of Feynman diagrams. It also discusses a pure S-matrix approach to particle dynamics.

Chew, G. F. *The analytic S-matrix.* Benjamin, 1966. A very physical discussion of the pure S-matrix approach to dynamics.

Collins, P. D. B. and Squires, E. J. *Regge poles in particle physics.* Springer, 1968. This is a comprehensive and thorough treatment of all aspects of Regge Pole Theory, starting with potential theory and ending with applications to high-energy scattering of elementary particles.

Barger, V. D. and Cline, D. B. *Phenomenological theories of high energy scattering.* Benjamin, 1969. Presents the main features of scattering data for strong interactions and surveys theoretical attempts to understand them.

Burhop, E. H. S. (ed.). *High energy physics.* Vols. I–V. Academic Press. This is a multi-author work covering a wide area of high-energy physics. It is principally useful as a reference work. Of particular value are the articles on the K-Meson–Nucleon Interaction by B. H. Bransden (Vol. III, 1969) and a vast survey of the 'Elementary Particles with Strong Interactions by D. H. Miller (Vol. IV, 1969). 'Photo- and Electroproduction Processes', by A. Donnachie (Vol. V, 1972), is probably the most comprehensive treatment of this subject in the literature.

Marshak, R. E. Riazuddin and Ryan, C. D. *Theory of weak interactions in particle physics*. Wiley-Interscience, 1969. A comprehensive treatise with long introductions on both physical and mathematical topics.

Konopinski, E. J. *The theory of beta radio-activity*. Clarendon, 1966. A beautifully clear exposition of the physical ideas involved in weak interactions, and of the historic experiments.

Carruthers, P. A. *Spin and isospin in particle physics*. Gordon and Breach, 1971. A pedagogical catalogue of the properties of the Lorentz, Poincaré and Orthogonal Groups and their role in particle physics. Gives the modern interpretation of the transformation properties of fields and relativistic wave equations.

Adler, S. L. and Dashen, R. *Current algebra*. Benjamin, 1968.

Renner, B. *Current algebras and their applications*. Pergamon, 1968.

Both these texts give an excellent treatment of the fundamental ideas of current algebra and of the many and widespread applications.

Kokkedee, J. J. J. *The quark model*. Benjamin, 1969. A collection of papers on many aspects of the quark model, with additional explanatory chapters.

Lee, B. W. *Chiral dynamics*. Gordon and Breach, 1972. A very specialised monograph, dealing with one of the more exciting offshoots of current algebra.

MATHEMATICAL TEXTS AND TABLES

Up to the beginning of this century, it very often happened that a major development in mathematics was stimulated by an unsolved problem in physics. With the massive growth in all scientific activity in the twentieth century, pure mathematics has largely divorced itself from physics and has proceeded along many and diverse abstract paths; although, from time to time, there is still a fruitful co-operation between the disciplines, viz. Dirac's δ function and the Theory of Distributions. As a consequence, the mathematical tools used by most physicists lag several years, or even decades, behind the latest mathematical developments, and a physicist will find himself quite lost if he consults the modern mathematical literature. (I am somewhat heartened to learn from a colleague that most *mathematicians* experience the same sense of bewilderment outside their own speciality!) One has to rely, therefore, on the occasional appearance of monographs that deliberately set out to teach these new branches of mathematics to the physicist.

For a general introduction to the mathematical methods of physics, we would suggest

Mathematical methods of physics, by J. Mathews and R. Walker (Benjamin, 1965).

A more advanced treatise that looks in greater depth at many of the topics introduced in the above volume, and which could follow it, is *Methods of mathematical physics*, by R. Courant and D. Hilbert Interscience, 1953). Volume I deals with integral equations, eigenvalue problems and calculus of variations. Volume II is devoted to partial differential equations.

Somewhat more specialised, but still excellent as introductory texts, are the following:

Titchmarsh, E. C. *The theory of functions*. 2nd edition. OUP, 1939. Treats basic complex variable theory.

Whittaker, E. T. and Watson, G. N. *A course on modern analysis*. 4th edition. CUP, 1927. A standard reference on the applications of analysis.

Hamermesh, M. *Group theory*. Addison-Wesley, 1962. A lucid and thorough exposition of group theory with many examples of its applications to physics.

Smithies, F. *Integral equations*. CUP, 1958. A short, beautifully clear account of the essential points; especially good on Fredholm Theorems.

At the very specialised level one may consult:

Herman, R. *Lie groups for physicists*. Benjamin, 1966.

Herman, R. *Vector bundles in mathematical physics*. Benjamin, 1970.

Naimark, M. A. *Linear representations of the Lorentz group*. Pergamon Press, 1964

Rühl, W. *The Lorentz group and harmonic analysis*. Benjamin, 1970. Detailed mathematical discussion of representations and expansion theorems.

Turning, now, to more specifically reference-type works, we shall list only a very few of the most useful and helpful volumes dealing with properties of functions, integral transforms and tables of integrals.

The most detailed and comprehensive discussion of the properties of the transcendental functions is given in:

Higher transcendental functions. Vols. 1–3. The Bateman Manuscript Project directed by A. Erdélyi. McGraw-Hill, 1953. Many of the results are derived in the text with great economy. This is a vast storehouse of knowledge—and it is easy to use.

Tables of integral transforms. Vols. 1 and 2. The Bateman Manuscript Project directed by A. Erdélyi. McGraw-Hill, 1954. Contains an extensive list of Fourier, Laplace, Mellin, Hankel, Stieltjes and Hilbert transforms, and others, as well as integrals of the higher transcendental functions.

For the more elementary properties of the special functions, as well as for much graphical and numerical data, one can refer to:

Jahnke, E. and Emde, F. *Tables of functions.* Dover, 1945. A pocket-sized volume.

Abramowitz, M. and Segun, A (eds.). *Handbook of mathematical functions.* Dover, 1965. A larger, and much more comprehensive, work. It includes some pedagogical material and examples to illustrate the use of the numerical tables. The approach is manifestly more modern than in the Jahnke–Emde volume.

An excellent reference work for tables of definite and indefinite integrals is:

Gröbner, W. and Hofreiter, N. *Integraltafel.* Vols. I+II. Springer, 1961. Although the text is in German, the tables are very efficiently laid out and easy to use.

An even more comprehensive list of integrals can be found in:

Gradshteyn, I. S. and Ryzhik, I. M. *Tables of integrals, series and products.* Academic Press, 1965.

SUPPLEMENTARY LIST

General relativity

Synge, J. L. *Relativity: the general theory.* North-Holland and Interscience, 1960
Synge, J. L. *Relativity: the special theory.* 2nd edition. North-Holland and Interscience, 1965
Møller, C. The theory of relativity. 2nd edition. Clarendon, 1972
Bergmann, P. G. *Introduction to the theory of relativity.* Prentice-Hall, 1942 (covers general, special and unified field theory and has long been used by students, but unfortunately it is now out of print)

Statistical mechanics

Tolman, R. C. *The principles of statistical mechanics.* OUP, 1938

Abrikosov, A. A., Gorkov, L. P. and Dzyaloshinskii, I. Ye. *Quantum field theoretical methods in statistical physics.* 2nd edition. Pergamon, 1965 (translated from Russian by D. E. Brown)
Stanley, E. *Introduction to phase transitions and critical phenomena.* Clarendon, 1971

Solid state

Ziman, J. M. *Principles of the theory of solids.* 2nd edition. CUP, 1970
Kittel, C. *Introduction to solid state physics.* 4th edition. Wiley, 1971
Haug, A. *Theoretical solid state physics.* 2 Vols. Pergamon, 1972
(Translation by H. S. W. Massey of *Theoretische Festkörperphysik*, Deuticke, 1970)
Solid state physics: advances in research and application. Edited by Ehrenreich, Seitz and Turnbull. 1– , 1955– . Academic Press (Vol. 27 = 1972)

9

Astrophysics

A. J. Meadows and J. G. O'Connor

INTRODUCTION

Two major branches of astronomy—'classical astronomy' and 'astrophysics'—can be distinguished. The former concerns itself with the positions and motions of astronomical objects; the latter, with their physical properties. The literature to be discussed here obviously falls under the second heading. Since, however, physical properties can only be observed when an object's position is known, and since, too, a physical property such as mass depends on a knowledge of motions, the distinction between the two types of literature cannot be drawn too strictly.

This word 'observation' brings up immediately a significant difference between astrophysics and most other branches of physics, which is reflected in their respective literatures. Whereas most physics is oriented towards laboratory experimentation, astrophysics depends on object-oriented observation. As a result, astrophysicists require special types of data compilation—such as star catalogues or charts—to identify and categorise objects of interest.

ABSTRACTING SERVICES

The main physics abstracting services also include some astrophysics. Of the three major abstracts journals—*Physics Abstracts*, *Bulletin Signalétique* and *Referativnyĭ Zhurnal*—the last has the best coverage of astrophysics, and the first is the most limited in this direction. But the most comprehensive source of abstracts in astrophysics is provided by a specialist publication—*Astronomy and Astrophysics Abstracts*. This is prepared, under the auspices of the IAU (International Astronomical Union), by the Astronomisches Rechen-Institut at

Heidelberg, and is published by Springer. It appears as two bound volumes a year, each covering the literature published during one half of the year. The volumes are produced within 6 months of the end of the half-year concerned. *Astronomy and Astrophysics Abstracts* has only appeared since 1968, when it replaced an earlier comprehensive abstracts series, the *Astronomischer Jahresbericht*, which had been issued in single annual volumes from 1899 onwards. The major difference between the two is that the older publication was in German, whereas its replacement is primarily in English.

Astronomy and Astrophysics Abstracts contains both an author index and a detailed subject index, so that the retrieval of individual articles is straightforward. The abstracts are also arranged into over 100 general subject categories, which makes it possible to scan the output of work in a given area rapidly. Together with its forerunner, this journal provides an excellent guide to most astrophysical literature published during the twentieth century. Nevertheless, there are certain fringe areas of astrophysics where coverage is not complete. For example, the literature of meteoritics is widely scattered, and not all of it finds a place in *Astronomy and Astrophysics Abstracts*. Here additional bibliographical aids—such as *Mineralogical Abstracts*—are necessary.

The most important such fringe area where further aids are required comes under the amorphous heading of 'space science'. Not only is this a very wide-ranging category: a significant part of the research output is published in report form rather than through the regular journals. NASA (National Aeronautics and Space Administration), fortunately, recognised the need for additional bibliographical tools from early on, and has for some years co-ordinated the publication of two guides to the space science literature (both containing material of astrophysical interest). Space science material published in journals or monographs is abstracted in *IAA* (*International Aerospace Abstracts*), which is published by the American Institute of Aeronautics and Astronautics, New York. The contents overlap considerably with those of *Astronomy and Astrophysics Abstracts*, and to that extent *IAA* is a less useful aid than the second abstracts journal, *STAR* (*Scientific, Technical and Aerospace Reports*), which provides a comprehensive survey of the report literature in space science. *STAR* is published by the Scientific and Technical Information Division of NASA, in Washington D.C., and contains a considerable amount of material not to be found readily elsewhere. Although the bulk of this is generated by NASA itself, an attempt is made to cover other relevant reports: for example, information on ESRO (European Space Research Organisation) reports is included.

The abstracts described so far appear in independent abstracts journals. Abstracts can also be found in certain specialised journals devoted mainly to the publication of original research. For example, issues of *Solar Physics* include abstracts of papers on solar research appearing in other journals; while the journal *Icarus* provides once a quarter a classified set of abstracts dealing with its own research topic—solar system research. Groups of abstracts on particular areas of astrophysics have occasionally also appeared in monograph form: a good example is provided by *Nuclear astrophysics: a bibliographical survey*, by B. Kuchowicz (Gordon and Breach, 1967).

Most work of astrophysical importance has been carried out in the twentieth century, but it occasionally becomes necessary to refer to literature of an earlier period. For the period up to 1880 there is a ready (though not exhaustive) source of references in *Bibliographie générale de l'astronomie jusqu'en 1880*, by J. C. Houzeau and A. Lancaster (new edition in 3 vols., Holland Press, 1964). Between 1880 and 1899, when the *Astronomischer Jahresbericht* began to appear, there is something of a hiatus as regards information sources, although a bibliography for these two decades is available on microfilm (from University Microfilms Ltd).

If the older information sought was originally published in one of the major astronomical journals, another useful guide is provided by the cumulative indexes which most of these publish from time to time. For example, the *Astrophysical Journal* has for many years contained a high proportion of significant astrophysical research: its last cumulative index appeared in 1966, covering the period from 1962.

DATA COMPILATIONS

The most widely used source of astrophysical data is probably *Astrophysical Quantities*, by C. W. Allen (Athlone, 1973). 3rd edn. Another useful source of astronomical data—also a little out of date now—is provided by Landolt-Börnstein (New series). The volume labelled Group VI/I (Springer, 1965) covers astronomy and astrophysics (see p. 43).

A variety of sources of more specialised astrophysical data exists— most of them operating on a continuing basis, often with support from the IAU. Data concerning solar activity and solar–terrestrial relations are especially well organised. The *Quarterly Bulletin on Solar Activity* (produced at the Zürich Observatory) collates observations made at over 100 observatories; while more rapid, or more detailed, information on solar activity can be obtained from one of the World Data Centres (e.g. at Boulder, Colorado). An international network of these centres was established for the IGY (International Geophysical Year) in 1957–58, and has been maintained ever since: mainly, of course, to collect geophysical data.

The main data bank for information concerning the properties of stars is the Stellar Data Centre at the University of Strasbourg Observatory. This collects data from a number of observatories, and aims at providing a comprehensive store of information on the astrometric, photometric and spectroscopic properties of stars. The Centre issues an information bulletin, and—as is increasingly the case in astrophysics—the data can be provided in machine-readable form.

We should note finally that much astrophysical spectroscopy is dependent on laboratory work. Hence, astrophysicists frequently need access to fundamental spectroscopic data. Two important compilations in this field are *The solar spectrum 2935Å to 8770Å* by C. E. Moore, M. G. J. Minnaert and J. Houtgast (1966) and *Atomic energy levels*, by C. E. Moore (1949–58; 3 vols., with later extensions). Both of these works are published by the National Bureau of Standards, Washington, and it is to the publications of this body that first reference is usually made by astrophysicists.

EPHEMERIDES

Compilations of this type provide information—usually for a calendar year at a time—on such matters as the precise position of the Moon, Sun and planets throughout the year. Although not in themselves of importance in astrophysics, ephemerides provide some of the basic data needed in formulating an observing programme. Various countries issue ephemerides, but since there is considerable overlap in contents, the calculations are frequently carried out in collaboration. In particular, the UK and the USA co-operate to produce their ephemerides. The resultant volume—called in this country the *Astronomical Ephemeris*—is prepared at the Royal Greenwich Observatory and published annually, well in advance, by HMSO. The material contained in the *Astronomical Ephemeris* is also available in machine-readable form (information on this can be obtained from the Bureau International d'Information sur les Ephémérides Astronomiques in Paris).

VISUAL AIDS IN ASTROPHYSICS

The observational bias of astrophysics leads to a particular emphasis on visual material. As with ephemerides, some of this material is not of direct astrophysical interest, but may be used in devising an observational programme. However, some star charts give astrophysical data directly. For example, the *Atlas borealis 1950.0* (most easily obtained from the Sky Publishing Co., Cambridge, Mass., 1962) not only provides information on the positions of stars north of $+30°$ declination and brighter than 9.35 magnitudes: it also indicates their spectral types. (The series of star charts prepared in Czechoslovakia, of which *Atlas borealis* is one, are some of the most useful for astrophysicists working on the brighter stars.)

In general, the most significant star charts for astrophysicists are those which consist of direct photographs. A prime example of this type of chart is the *Palomar sky survey* (produced by the National Geographic Society—Palomar Observatory Sky Survey, Pasadena, Calif.). In its basic form this consists of 935 pairs of photographic negatives (one set taken through a blue filter and the other through a red filter) going down to fainter than 20th magnitude. The Palomar sky survey only goes down to $-33°$ south: an extension to cover the rest of the southern hemisphere is now in preparation.

The use of direct photographs extends beyond whole-sky surveys: astrophysicists also need to refer to atlases of specific types of astronomical object. The astrophysical interest of such atlases is that morphology is frequently linked with physical properties. Photographic atlases appear in all branches of astrophysics: the following may be taken as typical examples of the various genres—the Lunar Orbiter photographs of the Moon's surface (see *Guide to Lunar Orbiter photographs*, by T. P. Hansen, NASA SP-242, 1970); the Mariner photographs of the Martian surface (*The Mariner 6 and 7 pictures of Mars*, by S. A. Collins, NASA SP-263, 1971); pictures of comets (*Atlas of cometary forms*, by J. Rahe, B. Donn and K. Wurm, NASA SP-198); photographs of normal galaxies (*The Hubble atlas of galaxies*, by A. Sandage, Carnegie Institution, Washington D.C., 1961) and of abnormal galaxies (*Atlas of peculiar galaxies*, by H. Arp, California Institute of Technology, Pasadena, Calif., 1966). Many of these photographic atlases are provided either boxed or in looseleaf form.

CATALOGUES

A catalogue lists astronomical objects, usually in order of their position in the sky, and provides information on one, or more, of their properties. The major catalogues are obviously important in astrophysics as repositories of astrophysical data; but knowledge of them is also required because they provide the designations by which astronomical objects are commonly known. For example, many stars are known by their HD number: that is by their position in the *Henry Draper Catalogue* (*Annals Astron. Observ. Harvard College*, *91*, 1918–99, 1924, with extensions in 100, 1925–36, and 112, 1949). Similarly, many galaxies and star clusters are known by their NGC number: that is by their position in the *New General Catalogue* (*New general catalogue of nebulae and clusters of stars* by J. L. E. Dreyer—republished with extensions by the Royal Astronomical Society in 1953).

Unfortunately, there are many catalogues now in existence. This leads not only to difficulty in tracking down particular items of information, but also to confusion in nomenclature, since a given object may appear in several different catalogues and so acquire a designation from each. Some of the most useful catalogues therefore collect both data and designations from a variety of sources and collate this material into a single list. So far as stars are concerned, two sources of this type may be singled out. The first is the *Catalogue of Bright Stars*, by D. Hoffleit (Yale University Observatory, New Haven, Connecticut, 1964). This provides comprehensive data (up to the date of publication) on stars down to about the limit of naked-eye observations. The second is the Smithsonian Astrophysical Observatory *Star catalog* (in four volumes, Smithsonian Institution, Washington D.C., 1966), which goes down to much fainter magnitudes. This Smithsonian catalogue is accompanied by a star atlas (obtainable from MIT Press, 1969).

Stars that show some characteristics in common may be grouped together in special catalogues, and designated in terms of these. For example, observations of visual binary stars have frequently been published in catalogue form. A major double star catalogue, such as *New general catalogue of double stars within 120° of the North pole*, by R. G. Aitken (Carnegie Institution, Washington D.C., 1932; 2 vols.), has provided the designation for numerous stars: in this case denoted by the letters ADS (Aitken Double Stars) plus a number. On the other hand, another major specialised catalogue, dealing with stars varying in their light output—*Obshchii katalog peremennikh zvezd*, by B. V. Kukarkin *et al.* (Akademiya Nauk, Moscow, 1958; 2 vols., with various extensions and supplements)—despite its importance is not a source of designations, since there is a special system of nomenclature for variable stars. This catalogue—usually referred to by its English title as the *General catalogue of variable stars*—is relatively unusual in that it forms part of an on-going project, so that its contents are up-dated from time to time. Like most Soviet publications, the catalogue and its supplements are not always easy to obtain.

Major specialised catalogues normally appear as free-standing monographs, but many smaller specialised catalogues are published as journal contributions. For example, standard lists of Be stars (i.e. stars with B-type spectra, but with emission lines) have appeared over a number of years, mainly in the *Astrophysical Journal*. Catalogues of this type can be difficult to track down: the safest approach is to find the standard monograph on the subject, which will usually refer to the important lists in journals. In the case of Be stars, for example, the standard text on this topic by A. B. Underhill, *The early type stars* (Reidel, 1966), contains this information.

General and specialised catalogues naturally exist not only for stars, but also for other kinds of astronomical object. These may appear in monograph form like most star catalogues. For example, the equivalent for galaxies of the *Bright star catalogue* is *Reference catalogue of bright galaxies*, by G. and A. de Vaucouleurs (University of Texas Press, 1964), which is similarly published in book form. But other forms of publication have been used. Thus, the most important catalogue of star clusters, *Catalogue of star clusters and associations*, by G. Alter, J. Ruprecht and V. Vanysek (Plenum, 1958—and frequently up-dated), is available solely as a boxed set of index cards. This has the advantage that the discovery of a new cluster only requires the insertion of an extra card in the sequence.

JOURNALS

Original research in astrophysics, as in other areas of physics, typically makes its first appearance in journals. Some papers of astrophysical interest can be found in general science journals such as the *Proceedings of the Royal Society* (Series A), but are concentrated mainly in rapid-publication journals of this type—especially *Nature* in the UK and *Science* in the USA. Astrophysical papers may also appear in the main-line physics journals (e.g. *Physical Review*), but the proportion is usually small unless the branch of physics dealt with overlaps an area of interest to astrophysics. Such overlap occurs most frequently in journals dealing primarily with instrumentation: for example, *Applied Optics* often contains articles relating in some way to astrophysics. Otherwise, physics journals normally only contain numerous articles of astrophysical interest when they are published in a country which has no journal of its own devoted to astronomy. Thus, a fair amount of the astrophysical research in Australia appears in the *Australian Journal of Physics*, since there is no major Australian astronomical journal.

In fact, most countries involved to any extent in astrophysical research do have one journal devoted specifically to astronomy and astrophysics, although only the larger countries have more than one major journal. In Czechoslovakia, for example, astrophysical research appears in the *Bulletin of the Astronomical Institutes of Czechoslovakia*, while the corresponding journal in Poland is *Acta Astronomica*. The papers in both these national journals are normally written in English, and this is true of much astrophysical research all over the world. Thus, the main Japanese journal dealing with astrophysics, *Publications of the Astronomical Society of Japan*, is published almost entirely in English.

English is, by a very wide margin, the favoured language for communication in astrophysics, although a country such as Germany naturally produces literature in its own language. Even here, however, one can see a trend towards English. An important development in recent years has been the creation of a European astronomical journal, formed from the amalgamation of the leading astronomical journals of France, Germany and the Netherlands. This new journal, *Astronomy and Astrophysics* (Springer, 1969–), accepts articles written in English, French or German, but the most commonly used language is certainly English.

Of all the major producers of astrophysical research, the Soviet Union makes least use of English in its communications. A number of journals, either partially or totally devoted to astrophysics, are produced in the USSR, and all are in Russian (or in the language of the republic concerned), but most of them include a title and an abstract in English. More importantly, most of the significant journals appear as cover-to-cover translations in the USA. Thus, the leading Soviet journal (*Astronomicheskii Zhurnal*) appears as *Soviet Astronomy AJ* (Consultants Bureau, N.Y.). *Astrofizika* (*Astrophysics*), *Astronomicheskii Vestnik* (*Solar System Research*) and *Kosmicheskie Issledovania* (*Cosmis Research*) are also available as cover-to-cover translations, although the delay time between the appearance of the Russian-language journal and the English translation is usually appreciably greater for these than for *Soviet Astronomy AJ*.

The United States currently produces some of the most significant astrophysical research. Correspondingly, the leading astrophysical journal in the world—the *Astrophysical Journal*—is published there (by the University of Chicago Press). The *Astrophysical Journal* has a correspondingly rapid-publication journal for short communications. (Called *Astrophysical Journal Letters*, this must be distinguished from *Astrophysical Letters*, which is a similar type of journal but published commercially by Gordon and Breach.) There is also a supplement series for the publication of longer papers, usually containing large accumulations of data.

Astrophysical Journal plays a role in astrophysical research comparable with that played by *Physical Review* in some other areas of physics. But it is by no means the only journal of astrophysical significance produced in the United States. The *Astronomical Journal* is more concerned with classical astronomy, but includes articles of astrophysical interest, especially on radio and radar astronomy. The *Publications of the Astronomical Society of the Pacific* published papers, usually fairly short, over the whole field of astronomy and astrophysics.

The main astronomical journal in the UK, the *Monthly Notices of the Royal Astronomical Society*, includes papers on both astrophysics and classical astronomy, but the former predominate. A rapid-publication letters section has recently been started, but it is included as part of the main journal. Longer papers, especially those including much numerical data, are published in the *Memoirs*. Another British journal, *The Observatory*, edited from from the Royal Greenwich Observatory, contains shorter notes often on astrophysical topics. It may be noted that Australian, Canadian, Indian and South African astronomers often publish their research in British or United States journals.

The journals mentioned so far are mainly published either by, or with the support of, astronomical societies. They have typically been in existence for many years—often dating back to the nineteenth century. As in other areas of physics, there has been a rapid increase in the number of astrophysical journals since World War II, but most of these have been produced on the initiative of commercial publishers. These are sometimes general in their contents, as the society journals are. For example, *Astrophysics and Space Science* (Reidel) covers virtually every branch of astrophysics. But more often a commercial journal concentrates on a specific branch of astrophysics, usually defined in terms of the objects studied. Thus, *Icarus* (Academic Press) and *Planetary and Space Science* (Pergamon) deal with solar system studies, as does the rapid-publication journal *Earth and Planetary Science Letters* (North-Holland). Many commercial journals are even more specific than this, as their titles indicate: examples are *Solar Physics* and the *Moon* (both published by Reidel) and *Physics of the Earth and Planetary Interiors* (North-Holland). Two points may be noted from these various titles. First, the commercial journals are particularly concerned with the physics of the solar system—a result of the growth of the space sciences during the 1960s. Second, a few commercial publishers produce most of the journals. Despite this latter point, and the overwhelming predominance of English as the language employed, these journals are designed to be international in both readership and authorship.

Solar system research, and space science in general, overlaps considerably from astrophysics into other areas of research. A full coverage of the literature therefore requires the inclusion of journals whose main emphasis would not normally be considered astrophysical. For example, anyone working on the physics of the solar system would need to consult the *Journal of Geophysical Research* and *Geochimica et Cosmochimica Acta*, the former a society publication (of the American Geophysical Union) and the latter a commercial publication (Pergamon).

REVIEW JOURNALS

The primary purpose of the journals described in the preceding section is the publication of original research, although some of them, from time to time, may publish review articles. However, a number of journals exist whose primary aim is to publish reviews of astrophysics. A few of these are published by learned societies (e.g. the *Quarterly Journal of the Royal Astronomical Society*), but the majority are commercial undertakings. As with the research journals, some physics review journals publish articles on astrophysics. An important example is *Reports on Progress in Physics* (London).

Review publications can be conveniently divided into two groups depending on whether they appear more frequently than once a year, or not. More frequent publications are normally produced in ordinary journal format, whereas volumes appearing at yearly or greater intervals generally resemble bound monographs. The former group of review journals mainly refer to areas of astrophysics that are involved in some way with observations from space—another consequence of the rapid growth of the research literature in this field during the last few years. Two useful journals of this type are *Space Science Reviews* (Reidel) and *Reviews of Geophysics and Space Science* (American Geophysical Union). In a slightly different category comes *Comments in Astrophysics and Space Science* (Gordon and Breach), which aims at commenting critically on current controversial points in astrophysics rather than reviewing them in depth.

Annual volumes of reviews often have a wider scope, covering the whole of astronomy and astrophysics. Within this group we have *Annual Reviews of Astronomy and Astrophysics* (some reviews of astrophysical interest appear in other volumes of this series, but are, in any case, referred to in the *Astronomy and Astrophysics* volume), *Advances in Astronomy and Astrophysics* (Academic Press) and *Vistas in Astronomy* (Pergamon), although this last title has not always appeared at strictly annual intervals. (Similar publication irregularities can occur in the production schedule of conference proceedings, even if these are supposed to be held at regular intervals.)

Most conference proceedings are a mixture of reviews and original research papers. Some appear more or less annually (e.g. *Space Research*, which stems from COSPAR—Committee on Space Research —meetings), but others appear, on average, less frequently than once a year: the *Colloque International d'Astrophysique, Liège*, for example. The *Transactions of the International Astronomical Union* form a special case, since, although they appear once every 3 years, they are published before each conference, and consist almost entirely of reviews. Some conference proceedings are, of course, one-off affairs, but most of the astrophysically interesting volumes form part of a continuing sequence. In some instances only occasional volumes in the sequence possess an astrophysical interest: this is true, for example, of the proceedings of the Enrico Fermi International School of Physics, organised by the Società Italiana di Fisica. Some sequences of conferences are only linked by the fact that they have the same sponsorship. The major funding agencies for astrophysical conferences in Europe and the USA include NASA, ESRO and NATO, but whereas the former two publish many of these conferences in their own report series (see below) NATO conference proceedings are published by a variety of independent publishers.

The IAU-sponsored symposia and colloquia deserve a paragraph to themselves. They occur frequently—perhaps several in a year, though at no fixed interval of time apart—and are each devoted to one specific topic. The resulting volumes contain mixed review and research material, and often represent the most comprehensive discussion of the topic available. IAU symposia may thus represent the definitive work on a subject and, as such, some of them will be mentioned below under the heading of monographs.

Owing to the increased cost of book publication and, perhaps equally important, the extra time-lag involved, it is becoming increasingly popular to publish the proceedings of a conference in one of the regular journals. They are particularly likely to appear in commercially produced journals, since topic-oriented conferences may link in readily with subject-oriented journals.

REPORT PUBLICATIONS

The largest producer of report literature in astrophysics is NASA. This organisation generates various reports series, some of little astrophysical interest. The literature of most relevance appears in the series labelled NASA-SP (standing for 'Special Publications'), including conference proceedings, data compilations and visual material such as the Lunar Orbiter photographs. ESRO also produces a report series, though on a much smaller scale than NASA, and this includes both review articles and original research. These two agencies provide most of the official report literature in astrophysics, although occasional publications of other government agencies (e.g. reports produced by RAE Farnborough) may be of some relevance.

Astrophysics appears, however, in another type of astronomical report of widespread importance—the observatory publication. These observatory reports usually consist of original research carried out wholly or partly by members of the institution concerned, though possibly funded by an external agency. The most important publications of this type in the UK are the *Royal Observatory Bulletins* (published by HMSO for the Royal Greenwich Observatory), followed by the *Publications of the Royal Observatory, Edinburgh* (also published by HMSO). Although various other observatories in the UK have produced their own report series, there seems now to be a trend for research material to be published preferentially in the regular journals. Correspondingly, some observatory reports have either been discontinued or been continued as a reprint series, in recent years. A similar trend can be discerned in North America, although a number of major observatories still produce their own publications. Important examples include the *Lowell Observatory Bulletin* (Flagstaff, Arizona) and the *Smithsonian Astrophysical Observatory Special Reports* (Cambridge, Mass.) in the United States, and the *Publications of the David Dunlap Observatory* (University of Toronto) and the *Publications of the Dominion Astrophysical Observatory* (Victoria, B.C.) in Canada.

In other areas of the world observatory reports frequently figure largely in the output of astronomical literature. They are, for example, a common form of publication in the USSR. One of the most important there is *Izvestia Krimskoi Astrofizicheskoi Observatorii* (Moscow), which counts as one of the major Soviet astrophysical journals.

The observatory report literature must be regarded, in bibliographical terms, as very poorly organised. Titles of report series have often changed; it is frequently uncertain whether, or not, a particular series is still appearing; and there may well be difficulty in obtaining copies. A recent publication by Rosenberg and Jansen, *Bibliography of non-commercial publications of observatories and astronomical societies* (contact the Utrecht Astronomical Institute), does provide some guide to this literature, and can be referred to in case of doubt.

MONOGRAPH LITERATURE

A discussion of the monograph literature of astrophysics cannot, of course, be in any sense exhaustive. We will concentrate here mainly on recently published monographs which contain good summaries and/or good bibliographies. Astronomical literature is nearly always classified on a straightforward object-oriented basis (*vide* the Dewey decimal system), and we shall follow the same pattern here.

Solar system

We may note first of all certain general volumes which discuss several objects in the solar system simultaneously. Perhaps still the most important of these is that edited by B. M. Middlehurst and G. P. Kuiper, *The solar system* (University of Chicago Press; published in 4 vols. between 1953 and 1963). These volumes are now distinctly dated—especially the first two, dealing with the Earth and the Sun—but they still provide interesting insights. A lower-level text, but more recent, well-written and with an excellent bibliography is *Moons and planets*, by W. K. Hartmann (Bogden and Quigley, N.Y., 1972). A more specialised volume dealing with the solid planet is *Surfaces and interiors of planets and satellites*, edited by A. Dollfus (Academic Press, 1970), while atmospheres are covered in *Planetary atmospheres*, edited by C. Sagan, T. C. Owen and H. J. Smith (Reidel, 1971; IAU Symposium No. 40). Another volume dealing with planets in general is *Physics of the solar system*, edited by S. I. Rasool (NASA SP-300, 1972).

The Moon is dealt with in a volume of that title edited by S. K. Runcorn and H. C. Urey (Reidel, 1972; IAU Symposium No. 47). A slightly older work, but with a more systematic coverage, is *Physics and astronomy of the Moon*, edited by Z. Kopal (Academic Press, 1971). Lunar studies have been changing with great rapidity in recent years: this is well reflected by the conferences on returned lunar materials. The research developments here are reported in *Proceedings of the Apollo 11 Lunar Science Conference* (Pergamon Press, 1970); *Proceedings of the Second Lunar Science Conference* (MIT Press, 1971); *Proceedings of the Third Lunar Science Conference* (MIT Press, 1972). All of these were published in sets of three volumes, and labelled as supplements to the journal *Geochimica and Cosmochimica Acta*.

The same rapid progress is evident in studies of Mars, but here no really up-to-date monograph exists, although S. Glasstone's

The book of Mars (NASA SP-179, 1968) provides a good summary of the older literature. In a similar way, no sufficiently recent monograph on Venus exists; and, although W. Sandner's *The planet Mercury* (Faber, 1963) is still moderately helpful, the planned space programme may soon drastically alter the state of our knowledge concerning that planet. Among the major planets, Jupiter, Saturn and Uranus have all had monographs devoted to them—respectively, by B. M. Peek, *The planet Jupiter* (1958); by A. F. O'D. Alexander, *The planet Saturn* (1962); and by the same author, *The planet Uranus* (1965). All three are published by Faber and Faber. The volume on Jupiter is now seriously out of date and must be supplemented by more recent review articles. The other two surveys, although requiring some updating, are less affected. Neptune and Pluto have not yet received monographic treatment.

Apart from the planets, the solar system contains various smaller bodies. Asteroids and comets have both been covered by recent conference proceedings: in the former instance edited by T. Gehrels, *Physical studies of minor planets* (NASA SP-267, 1972; IAU Colloquium No. 12); in the latter, by G. P. Kuiper and E. Roemer, *Comets—scientific data and missions* (Lunar and Planetary Laboratory, Tucson, Ariz., 1972). G. J. McCall's *Meteorites and their origins* (David and Charles, 1973) provides a brief bibliography of important work in this field. A general guide to the interplanetary medium, though a little out-of-date now, is *The Zodiacal light and the interplanetary medium* (NASA SP-150, 1967). The gas component is treated in *The solar wind*, edited by C. P. Sonett, P. J. Coleman Jr. and J. M. Wilcox (NASA SP-308, 1972), and its interaction with the Earth in a massive monograph by S.-I. Akasofu and S. Chapman, *Solar–terrestrial physics* (Clarendon, 1972).

The origin of the solar system has been discussed at various recent conferences—importantly at a symposium in France (Reeves, H., ed., *L'origiue du Système Solaire,* Paris, CNRS, 1974). The spacing of the planets in the solar system is discussed, with good bibliographical detail, in M. M. Nieto's *The Titius–Bode law of planetary distances* (Pergamon, 1972).

Numerous monographs have been published about the Sun, either concerning its properties in general or, more usually, devoted to some specific aspect. In the former category H. Zirin's *The solar atmosphere* (Blaisdell, 1966) still provides a useful, if sometimes idiosyncratic, introduction. The more specialised texts can be divided into those which deal with particular parts of the solar atmosphere and those which deal with solar activity. In the former group two monographs by R. J. Bray and R. E. Loughhead, *Sunspots* (Chapman and Hall, 1964) and *The solar granulation* (Chapman and Hall, 1967),

can be used for preliminary reading, although both are now somewhat out of date. Another text by the same authors is *The solar
chromosphere* (Chapman and Hall, 1974). Studies of the corona have
been well covered in a fairly recent volume: *Physics of the solar
corona*, edited by C. J. Macris (Reidel, 1971). Two fairly recent IAU
symposia provide between them a reasonably detailed introduction
to the study of solar activity—*Structure and development of solar
active regions*, edited by K. O. Kiepenheuer (Reidel, 1968; IAU
Symposium No. 35) and *Solar magnetic fields*, edited by R. Howard
(Reidel, 1971; IAU Symposium No. 43). A more unified, though
rather less detailed, treatment can be found in E. Tandberg-Hanssen's
Solar activity (Blaisdell, 1967). Solar radio observations are included
as an integral part of texts such as these, but the complex question of
radio bursts has recently been examined in a separate monograph by
A. Krüger, *Physics of solar continuum radio bursts* (Akademie Verlag,
Berlin, 1972).

Space physics does not really count as a separate branch of astrophysics in terms of objects studied: it is mainly concerned with the
areas we have already described. Nevertheless, not only do the
techniques of space physics differ from those used in ground-based
work, but also certain observations (e.g. of the solar wind) are made
almost exclusively by satellite. It is therefore worth noting that
several books devoted to space physics have appeared in recent years.
A title that can be recommended is by R. C. Haymes, *Introduction to
space science* (Wiley, 1971).

Stars and galaxies

As with the solar system, we must first note a series of volumes,
again with G. P. Kuiper and B. M. Middlehurst as general editors.
Under the over-all title *Stars and stellar systems*, the series provides an
excellent introduction to astrophysics outside the solar system,
including both instrumentation and techniques. This nine-volume
series, published by the University of Chicago Press, has appeared
over more than a decade and, as a result, some of the individual
articles are now out of date. So far as stellar astrophysics is concerned,
a fairly recent, shorter set of collected articles can be found in
Stellar astronomy, edited by H. Y. Chiu, R. L. Warasila and J. L. Remo
(Gordon and Breach, 1969; 2 vols.).

Stellar atmospheres are dealt with in a number of volumes; one
general text that can be recommended is by D. Mihalas, *Stellar
atmospheres* (Freeman, 1970). L. H. Aller's *Astrophysics* (Ronald,
1963; 2nd edition—first volume only), though now somewhat out-

dated, still offers excellent preliminary reading. In the field of stellar spectroscopy two volumes with that title by M. Hack and O. Struve (available from Osservatorio Astronomico di Trieste, 1969 and 1971, with possibly more volumes to follow) can be recommended for initial study. A recent detailed theoretical treatment of spectral lines and radiative transfer in stellar atmospheres can be found in R. G. Athay's *Radiation transport in spectral lines* (Reidel, 1972). General discussions of atomic and molecular spectra have appeared in the proceedings of a recent NATO Advanced Study Institute: *Atoms and molecules in astrophysics*, edited by T. R. Carson and M. J. Roberts (Academic Press, 1972).

Several volumes are also available dealing with stellar interiors. Two titles may be mentioned: J. P. Cox and R. T. Giuli's *Principles of stellar structure* (Gordon and Breach, 1968; 2 vols.) and *Stellar evolution*, edited by H.-Y. Chiu and A. Muriel (MIT, 1972).

We may link with stellar structure the study of astrophysical plasmas and magnetohydrodynamics: but here the situation is different, for, although several good review articles exist, there are few general monographs. J. H. Piddington's *Cosmic electrodynamics* (Wiley-Interscience, 1969) is now somewhat outdated, but is still a useful source for preliminary reading.

A good recent introductory text for the study of our own Galaxy is D. Mihalas and P. M. Routly's *Galactic astronomy* (Freeman, 1968). It can be supplemented by two volumes of collected articles: *Structure and evolution of the Galaxy*, edited by L. N. Mavridis (Reidel, 1971; the proceedings of a NATO Summer School) and *The spiral structure of our Galaxy*, edited by W. Becker and G. Contopoulos (Reidel, 1970; IAU Symposium No. 38). These volumes naturally include discussions of the interstellar medium, but a more systematic, if less up-to-date, account can be found in a text by S. A. Kaplan and S. B. Pikelner—translated from the Russian—*The interstellar medium* (Harvard University Press, 1970). One of the recent growth points in interstellar studies—the study of interstellar molecules—is mainly discussed in review articles; but a new volume, edited by M. A. Gordon and L. E. Snyder, is *Molecules in the galactic environment* (Wiley, 1973).

In terms of objects within our Galaxy, the prime source of information is often an IAU Symposium. Recent examples include *Planetary nebulae*, edited by D. E. Osterbrock and C. R. O'Dell (Reidel, 1968; IAU Symposium No. 34) and *The Crab nebula*, edited by R. D. Davies and F. G. Smith (Reidel, 1971; IAU Symposium No. 46). But major groups of stars, or ones currently at the centre of astrophysical interest, often have commercially produced volumes written about them. Thus, two important types that have recently formed the subjects of general review monographs are dealt with in W. Strohmeier's *Variable stars* (Pergamon, 1972) and A. H. Batten's *Binary and multiple systems of stars* (Pergamon, 1973). An example of a commercially produced text on an astronomical object currently at the focus of attention is provided by *The physics of pulsars*, edited by A. M. Lenchek (Gordon and Breach, 1972).

One of the most striking developments of recent years in astronomy has been the extension of astrophysical observations to parts of the spectrum other than the visual and the radio. Most of the monograph literature resulting from this advance actually consists of conference proceedings: one of the few genuine texts in this area that can be recommended for preliminary reading is T. C. Weekes's *High-energy astrophysics* (Chapman and Hall, 1969). Among the conference proceedings, IAU symposia again figure largely. Two that contain relatively up-to-date summaries are: *Ultraviolet stellar spectra and related ground-based observations*, edited by L. Houziaux and H. E. Butler (Reidel, 1970; IAU Symposium No. 36) and *X- and Gamma-ray astronomy*, edited by H. Bradt and R. Giacconi (Reidel, 1973; IAU Symposium No. 55).

We finally consider monographs dealing with external galaxies and with cosmology. A good introductory link between work in these areas and problems within our own Galaxy is provided by *Astrophysics and general relativity*, edited by M. Chrétien, S. Deser and J. Goldstein (Gordon and Breach, 1969; 2 vols.). Another general text on galaxies for preliminary reading is by J. Lequeux, *Structure and evolution of galaxies* (Gordon and Breach, 1969). Of recent conference proceedings, two cover areas of major current interest in studies of galaxies. The first is edited by D. J. K. O'Connell, *Nuclei of galaxies* (North-Holland, 1971); the second by D. S. Evans, *External galaxies and quasi-stellar objects* (Reidel, 1972; IAU Symposium No. 44). D. Sciama's *Modern cosmology* (CUP, 1971) is an excellent text for a preliminary study of cosmology, but it has few bibliographical references. However, S. Weinberg's *Gravitation and cosmology* (Wiley, 1972)—also an outstandingly good, if much more advanced, text—contains an extensive guide to the literature.

CONCLUDING REMARKS

Anyone studying the astrophysical literature up to the end of the 1960s should consult an excellent bibliography by D. A. Kemp, *Astronomy and astrophysics: a bibliographical guide* (Macdonald, Archon, 1970). It can be supplemented with regard to the work of the International Astronomical Union by reference to the *Astronomer's handbook*, edited by J.-C. Pecker (Academic Press, 1966; *Trans. IAU*, **XII C**).

Astrophysical literature is by no means so widely available in the UK as the literature in many other branches of physics: many university libraries, for example, have only an incomplete coverage. It may therefore be worth recording that the three main specialist libraries in the UK dealing with astrophysics are those of the Royal Astronomical Society in London, the Royal Greenwich Observatory in Sussex and the Royal Observatory, Edinburgh.

10

Mechanics and sound

R. H. de Vere

MECHANICS

Mechanics is that branch of science which deals with the effects of force upon bodies at rest or in motion. It is conveniently divided into the study of rigid bodies (i.e. dynamics, classical mechanics), of elastic bodies (i.e. solid mechanics) and of fluids (i.e. fluid mechanics). The combination of solid and fluid mechanics is often referred to as continuum mechanics. Since sound is a mechanical disturbance propagated through an elastic medium, it may quite naturally be treated in a chapter on mechanics. Similarly, vacuum techniques may be dealt with alongside fluid mechanics.

Because of its long and close association with engineering, it is not always easy in the literature to determine at which point mechanics ceases and engineering begins. Fortunately, there is an area of study—applied mechanics—which is sufficiently fundamental to interest the physicist while still being of technological interest. Many of the works discussed below fall into this category, although an attempt has been made to reject those where too great an emphasis is placed on engineering.

Mechanics is a well-established subject and many of the works written from the late nineteenth century onwards may be read with profit even now. However, only modern works are considered here.

Most of the published literature in mechanics is devoted to one or other of the sub-divisions outlined above, and are discussed in the relevant sections below. A modern theoretical treatment of all branches of the subject is to be found in the relevant volumes of *Encyclopedia of Physics*. Details of the separate volumes are given in the appropriate sections below (see also p. 37).

Review papers are to be found in *Acta Mechanica* (monthly;

Springer, 1965–); papers are in English or German, with summaries in either language. Other review articles and papers on recent developments are to be found in the periodical literature and the published proceedings of conferences. Examples of such all-embracing journals are *Annales scientifiques. 2. Série. Mécanique et Physique théorique* (Besançon Université, 1957–) and *Mechanics Research Communications*. Pergamon, 1974– . Bi-monthly. Sponsored by the International Society for Mechanical Sciences. An international journal for the rapid communication of contributions to the mechanics of fluids, solids, particles and systems.

Space limitations preclude a complete list of journal titles. Some titles have been rejected because they lack sufficient technical content; others for the reasons given below, in spite of their technical value.

1. Journals in difficult languages, such as Chinese, Japanese, Polish, etc., which contain some English in the form of summaries or lists of contents. Examples are *Japanese Journal of Experimental Mechanics,* and the Polish *Rosprawy mzynierskie, Mechanika Teoretyczna i Stosowana.*

2. Report literature published by a wide variety of institutions, such as the von Karman Institute for Fluid Dynamics in Belgium, Aeronautical Research Council (London), Tohuke University Institute for High Speed Mechanics, University of Southampton Department of Aeronautics and Astronautics. Such reports are not always published in the more formal literature, and frequently contain some very useful information—particularly data.

3. Journals whose engineering content is large but which nevertheless contain some fundamental mechanics. Examples are *Bulletins of the Japan Society of Mechanical Engineers, Journal of Applied Mechanics, Progress in Aeronautical Sciences, Soviet Aeronautics, Transactions of the Royal Institution of Naval Architects.*

Mechanics is abstracted by the major abstracting services, such as *Physics Abstracts, Physikalische Berichte, Engineering Abstracts Monthly*, etc., and it is to such publications one must turn when searching on a broad front.

An important service covering the whole range of mechanics is:

Applied Mechanics Reviews. American Society of Mechanical Engineers, 1948– . Monthly. 10 000 informative and indicative abstracts p.a. culled from some 400 periodicals, reports, conference proceedings and monographs. Annual indexes make this a useful tool for retrospective searching.

For translators a useful multi-lingual dictionary specialising in mechanics is:

Vocabulary of Mechanics in Five Languages; Polish, German, French, Russian and English. A. T. Proskolanski, editor. 2 vols. Vol. 1, parts 1 and 2, Pantswowe Wydawctwa Technczne, 1959, 1960. Vol. 2, Pergamon, 1967. Volume 1 covers theoretical mechanics and strength of materials and the terms are defined in Polish. Volume 2 covers fluid mechanics and the terms are defined in English.

Many associations exist at national and international level to promote and encourage the development of theoretical and applied mechanics. Their function is to co-ordinate the actions of smaller organisations, to establish working parties to study particular topics (such as glossaries, training or standardisation) or to instigate meetings. They may also publish news bulletins, congress proceedings, bibliographies, technical journals, etc. Examples of such organisations are: International Union of Theoretical and Applied Mechanics, International Committee on Rheology, Society for Experimental Stress Analysis, British Society of Rheology.

Dynamics

Dynamics is concerned with the motion of a system of particles under the influence of forces, especially those originating from outside the system. Treated under this heading are oscillators, vibrations and waves, and gyroscopic motion. Celestial mechanics is treated in the chapter on astrophysics. Among the general books on dynamics are:

Barford, N. C. *Mechanics.* Wiley, 1973

Cannon, R. H. *Dynamics of physical systems.* McGraw-Hill, 1967. Not restricted to mechanics, but it does underline the application of the methods of mechanics to other physical phenomena.

More advanced treatments are to be found in:

Pars, L. A. *A treatise on analytical dynamics.* Heinemann, 1965. A comprehensive account for the theoretician of dynamics as it now stands.

Saletin, E. J. and Cromer, A. H. *Theoretical mechanics.* Wiley, 1971. Attempts to present classical mechanics in such a way as to facilitate the study of modern quantum mechanics and field theory.

Oscillations, vibrations and waves

Oscillations, vibrations and waves occur in many branches of science other than mechanics. Here we deal with mechanical oscillations and waves on a macroscopic scale.

The theory of oscillators may be studied in:

Andronov, A. A., Vitt, A. A. and Kaikin, S. E. *Theory of oscillators*. Pergamon, 1966

Bishop, R. E. and Johnson, D. C. *The mechanics of vibration*. CUP, 1960

Haberman, C. M. *Vibration analysis*. Merrill, 1968. An introductory text, approaching the subject from the viewpoint of computer-assisted solutions.

Minorsky, N. *Nonlinear oscillators*. Van Nostrand. 1962

Volterra, E., and Zachmaoglou, E. C., *Dynamics of Vibration*. Merrill, 1965

Recent developments appear in the more general periodicals and conference proceedings, but one journal which devotes itself to the problems of dynamics and vibrations (including electrical, thermal and acoustical matters) is *Proceedings of Vibration Problems* (Institute of Basic Technical Problems of the Polish Academy of Sciences, 1959– . Quarterly). In English with Polish and Russian summaries.

Finally, two delightful books which illustrate the reality behind so much theory. They are *Chladni figures, a study in symmetry*, by M. D. Waller (Bell, 1961), a heavily illustrated book showing the many oscillatory modes on a vibrating plate; and *Cymatics. The structure and dynamics of waves and vibrations*, by H. Jenny (Moos, 1967), a beautifully illustrated record of waves, vibrations and periodicity occurring throughout nature.

The propagation of vibrations throughout a medium is well enough treated in most physics textbooks but there are some important aspects which receive special attention. Two works are mentioned here; other works on the propagation of waves in elastic media are discussed in the section on acoustics.

Brekhovsikh, L. M. *Waves in layered media*. Academic Press, 1960

Brillouin, L., and Porod, M. *Propagation des ondes dans les milieux périodiques*. Masson, 1956

Gyrodynamics

Gyroscope theory and application are well treated in a number of monographs, such as:

Arnold, R. N. and Maunder, L. *Gyrodynamics and its engineering applications*. Academic Press, 1961

Magnus, K. *Kreisel. Theorie und Anwendungen*. Springer, 1971

Current papers are to be found in periodicals covering a wide range of subjects. The papers are well covered in such abstracting services as *Applied Mechanics Reviews.*

Continuum mechanics

Continuum mechanics is a term coined to cover the underlying theories of mechanics applicable to both the solid and the fluid states. It is a sub-division of continuum physics (with which it is often confused) which is concerned with a wider range of phenomena such as electromagnetism, heat flow, etc. The term is also used to describe literature whose contents cover both solid and fluid mechanics. Substantial treatments are to be found in:

Eringin, A. C. *Continuum physics.* Academic Press, 1971. 3 vols.

Malvern, L. E. *Introduction to the mechanics of a continuous medium.* Prentice-Hall, 1969

Sedov, L. I. *A course in continuum mechanics.* Wolters-Noordhof. Vol. 1, *Basic equations and analytical techniques*, 1971. Vol. 2, *Physical foundations and formulation of problems*, 1972. Vol. 3, *Fluids, gases and the generation of thrust*, 1972. Vol. 4. *Elastic and plastic solids and the formation of cracks*, 1972 (contains index to all four volumes)

Critical reviews are to be found in *Advances in Applied Mechanics*, Academic Press, 1948– . This annual surveys the present state of research in various fields of applied mechanics. Other review articles are to be found along with original papers in the journal literature, such as:

Archives for Rational Mechanics and Analysis. Springer, 1957–. Irregular, four to five issues per volume. The papers—in English, French, German or Italian—are concerned with mathematical mechanics, promoting pure analysis particularly in the context of application.

Archives of Mechanics (formerly *Archivum Mechaniki Stoswanej*). Instytut Podstawowych Problemow Techniki, Polska Akademia Nauk, 1949–. Bi-monthly. Papers are mainly in English, with some in French.

International Journal of Engineering Science. Pergamon, 1963– . Monthly. Contains papers on original research on the application of physics, chemistry and mathematics to Engineering, considered from both the continuum and molecular points of view.

International Journal of Non-linear Mechanics. Pergamon, 1966– .

Bi-monthly. Papers, mainly in English but some in French, German or Russian; papers have abstracts in English, French, German and Russian. Contains research results of permanent interest and some invited review articles.

Journal of Applied Mathematics and Mechanics. Pergamon, 1958– . Bi-monthly. Cover-to-cover translation of *Prikladnaya Matematika i Mekhanika.* 22–, 1958– .

Journal de Mécanique. Gauthier-Villars (for CNRS), 1962– . Quarterly. In French; papers have brief summaries in English.

Journal of Applied Mechanics and Technical Physics. Consultants Bureau, 1966– . Bi-monthly. Cover-to-cover translation of *Zhurnal Prikladnoĭ Mekhaniki i Tekhnicheskoĭ Fiziki,* 1960– .

Meccanica, Italian Association of Theoretical and Applied Mechanics, 1966– . Quarterly (text in English).

Quarterly Journal of Mechanics and Applied Mathematics, Clarendon, 1948– .

Soviet Applied Mechanics. Faraday Press, 1968– . Monthly. Cover-to-cover translation of *Prikladnaya Mekhanika,* 1965– .

Recent developments are also reported in the proceedings of repeated conferences—local, national or international—such as *Congress on Theoretical and Applied Mechanics* (15th, Sindri, 1970). *International Congress of Applied Mechanics,* (12th, Stanford, 1968), *Japan National Congress for Applied Mechanics* (21st, Tokyo, 1971), *Midwest Conference on Mechanics,* (12th, Notre Dame, 1971). *South-east Conference on Theoretical and Applied Mechanics* (6th, Raleigh and Durham, 1972). *United States National Congress of Applied Mechanics* (6th, Harvard, 1970).

The specialist abstracting services for continuum mechanics include the following titles.

Current Awareness Bulletin for Mechanical Engineers. University of Surrey Library, 1970– . Fortnightly. Lists titles and references only in subject order. Covers applied mechanics quickly and efficiently.

Soviet Abstracts. Mechanics. Ministry of Technology, Technical Information and Library Service, 1956– . Monthly. Translations of abstracts selected from *Referativnyi Zhurnal. Mekhaniki.* About 2000 informative abstracts p.a. are selected from USSR and East European countries only. Reviews are flagged. There is a contents list in each issue but no indexes, which restricts its use to a current awareness service.

Solid mechanics

Solid mechanics (including elasticity, plasticity and rheology) is the study of the statics and dynamics of deformable bodies. It includes *elasticity*, which is concerned with substances which regain their original shape when relieved from stress; *plasticity*, which involves instantaneous (time-independent), permanent deformation of solids; and *rheology*, which is concerned with the time-dependent flow of quasi-solid materials. The behaviour of 'pure' fluids under stress is dealt with in the section on Fluid Mechanics.

Well-established texts suitable for a thorough grounding in the phenomenological theory of elasticity are to be found in *A treatise on the mathematical theory of elasticity*, by A. E. H. Love (CUP., 4th edn, 1927); *Elasticity fracture and flow*, by J. C. Jaeger (Methuen, 2nd edn, 1962); *Theory of elasticity*, by S. P. Timoshenko and J. N. Goodier (McGraw-Hill, 3rd edn, 1970); *Theoretical elasticity*, by A. E. Green and W. Zerna (Clarendon, 2nd edn, 1968); and *Theory of elasticity*, by L. D. Landau and E. M. Lifschitz (Pergamon, 1970).

An advanced exposition of the subject is given in *Encyclopedia of physics. Volume VI, Elasticity and plasticity* and *Volume VIa. Mechanics of solids* (see p. 37).

An up-to-date treatment of large deformations is given in A. E. Green and J. E. Adkin's *Large elastic deformations* (Clarendon, 2nd edn, 1970). Thermoelasticity is covered at an introductory level by H. Parkas's *Thermoelasticity* (Blaisdell, 1968); a more advanced treatment appears in *Thermoelasticity*, by W. Nowacki (Pergamon, 1962); and recent developments in *Trends in elasticity and thermoelasticity*, by R. E. Czarnota-Bojarski, M. Sokolowaki and H. Zorski (Wolters-Noordhof, 1971).

Plasticity is frequently dealt with as an adjunct to elasticity. *Foundations of the theory of plasticity*, by L. M. Kachanov (North-Holland, 1971), and *Dynamic plasticity*, by N. Cristescu (North-Holland, 1967) are two modern books. Recent developments are to be found in the periodical literature of theoretical and applied mechanics.

Periodicals restricted to solid mechanics include:

Engineering Fracture Mechanics. Pergamon, 1968–   . Quarterly

International Journal of Solids and Structures. Pergamon, 1965–   . monthly

Journal of the Mechanics and Physics of Solids. Pergamon. 1952–   . Bi-monthly. Aims to develop the fundamental ideas bearing on the connection between the continuum and micro-structural properties of materials.

Mechanics of Solids. Allerton Press Journal Program, 1969– . Bi-monthly. Cover-to-cover translation of *Mekhanika Tverdogo*, 1969– .

Experimental aspects of solid mechanics are well documented and include such monographs as *A treatise on photoelasticity*, by E. G. Coker, L. N. G. Filon and H. T. Jessop (CUP, 1957); *Moiré analysis of strain*, by A. J. Durelli and V. J. Parks (Prentice-Hall, 1970); and *Experimental strain analysis, principles and methods*, by G. S. Hollister (CUP, 1967). The latter work covers all modern methods.

Two journals in the experimental field are *Experimental Mechanics* (Society for Experimental Stress Analysis, 1961– . Monthly) and *Journal of Strain Analysis* (Institution of Mechanical Engineers, for the Joint British Committee for Stress Analysis, 1965– . Quarterly).

The major work on rheology is *Rheology. Theory and applications*, by F. R. Eirich (Academic Press, 1956–). Volume 5, 1969, is the most recent. This multi-volume treatise summarises the experimental and theoretical knowledge of the science of flow and deformation as related to physical and chemical properties. Other important works are *Elasticity, plasticity and structure of matter*, by R. Houwink and H. K. De Decker (CUP, 3rd edn, 1971); *Advanced rheology*, by M. Reiner (Lewis, 1971); and *Deformation, strain and flow. An elementary introduction to rheology*, by M. Reiner (Lewis, 3rd edn, 1969).

Two important journals are:
Rheologica Acta. Steinkopff, 1958– . Quarterly. Papers in English, French or German. Summaries in English, French and German are provided for each paper.
Transactions of the Society of Rheology. Interscience, 1957– . Quarterly. There is a cumulative index for 1957–66 in Vol. 10, pt. 2, 1966.

The *International Congress on Rheology* has met every 5 years since 1948, with the proceedings published a year or two later. The most recent congress was the 5th, Kyoto, 1968. The proceedings were published by University of Tokyo Press, 1970 (in English).

Rheology Abstracts (Pergamon, for the British Society of Rheology, 1958– . Quarterly) is the specialist service for the subject. It surveys the world's literature and gives both indicative and informative abstracts. Annual indexes make it a useful tool for retrospective searching.

Recent research activities in the United Kingdom are recorded in *Rheology research in Britain, 1970– . A survey*, by J. M. Hobbs and R. R. Martin (National Engineering Laboratory, NEL Report 493).

Tribology

Tribology is the science of contacting solid surfaces in relative motion, and is concerned with friction, lubrication and wear. Of great technological importance and scientific interest, its study embraces surface structure, adhesion, lubrication hydrodynamics and rolling friction. An idea of current activity in the field may be gleaned from *Critical reviews in tribology* (IPC Science and Technology Press, 1970) a review of the world's literature with 1263 references.

For an introduction to the subject, the following books are suggested: *Friction and wear of materials*, by E. Rabinowicz (Wiley, 1965); *Friction, lubrication and wear* by F. P. Bowden and D. Tabor (Methuen, 2nd edn, 1967); and *Tribology*, by E. D. Hondros (Mills and Boon, 1971). An advanced treatment is given in *The friction and lubrication of solids, part II*, by F. P. Bowden and D. Tabor (Clarendon, 1964). This is a sequel to part I (1954) and very much concerned with the surface phenomena of friction.

For translators there is *Glossary of terms and definitions in the field of friction, wear and lubrication tribology* (OECD, Research Group on Wear of Engineering Materials, 1969). In English, French, German, Spanish, Italian and Japanese.

Fluid mechanics

Fluid mechanics is concerned with the motion of matter in the gaseous, liquid and plasma states. At low velocities the theories of classical hydrodynamics suffice; at higher velocities compressible flow theory takes over and the flow is referred to as transonic, supersonic or hypersonic. When one deals with the applied branches of gas flow, e.g. air flow over aerofoil sections or through ventilator systems, the subject is called aerodynamics. When the medium is electrically conducting and in the presence of a magnetic field we refer to magneto-hydrodynamics (MHD). Superimposed on these major divisions are many important sub-divisions picked out for special treatment in the literature. Some of the sub-divisions are emphasised in this section, e.g. laminar flow, boundary layer flow, turbulent flow, constrained flow, multiphase flow, flow of rarefied gases, flow measurement, etc. Vacuum techniques are treated in the next section.

For a general introduction to classical fluid dynamics *Hydro-*

dynamics, by H. Lamb (CUP, 6th edn, 1932), is still to be recommended. A useful concentrated yet comprehensive survey of both theory and applications is to be found in *Handbook of fluid dynamics*, by V. L. Streeter (McGraw-Hill, 1961). For an expert exposition there is *Encyclopedia of Physics, Vols. VIII and IX. Fluid Dynamics* (see p. 37), a theoretical treatment of all aspects of fluid dynamics.

A few useful books, both introductory and advanced, are:

Brodkey, R. S. *The phenomena of fluid motion*. Addison-Wesley, 1967. A good introductory book.

Curle, N. and Davies, H. J. *Modern fluid dynamics*. Van Nostrand–Reinhold, 1961, 1967. Vol. 1. *Incompressible flow*. Vol. 2. *Compressible flow*.

Duncan, W. J., Thom, A. S. and Young, A. D. *Mechanics of fluids*, 2nd edn. Arnold, 1970. An introductory text.

Ladyzhenskaya, O. A. *The Mathematical theory of viscous incompressible flow*. 2nd edn. Gordon and Breach, 1969.

Lu, Pau-Chang. *Introduction to the mechanics of viscous fluids*. Holt, Rinehart and Winston, 1973

Skelland, A. H. P. *Non-Newtonian flow and heat transfer*. Wiley, 1967. An introductory text.

Yih Chia-Shun. *Fluid mechanics. A concise introduction to the theory*, McGraw-Hill, 1969

One of the most important developments in fluid mechanics is the concept of boundary layer flow. Recent works include *Non-isothermal laminar flow*, by P. B. Kwant (Uitgevers Maatschappij, 1971); *Boundary layer theory*, by H. Schlichting (McGraw-Hill, 6th edn, 1968); and *Boundary layers of flow and temperature*, by A. Walz (MIT Press, 1969).

Reviews appear in the *Annual Review of Fluid Mechanics* (Annual Reviews, 1969–). A short collection of reviews also appears in *Basic developments in fluid mechanics* (Maurice Holt, ed. Academic Press, 1965. 2 vols.

The periodical literature is extensive but there the selection is restricted to those journals which concentrate more on the fundamental aspects, such as:

Fluid Dynamics. Faraday Press, 1966– . Bi-monthly. (Cover-to-cover translation of *Izvestiya Mekhanika. Zhidosti i Gaza*, 1966– .

Fluid Mechanics. Soviet Research. Scripta Publishing, 1972– . Bi-monthly. English translations of selected Russian papers not otherwise available in English.

Journal of Fluid Mechanics. Taylor and Francis, 1956– . Twenty issues p.a.

Physics of Fluids. American Institute of Physics, 1958–   . Monthly.

The literature of the subject is abstracted by the usual abstracting services in physics and mechanics and also by such specialist services as:

Channel. A Current Information Guide. British Hydromechanics Research Association, 1968. Monthly. 1800 informative abstracts p.a. Semi-annual author and place name indexes make this useful for retrospective searching.

Fluid Mechanics Current Index. Fluid Mechanics Publications, 1972–   . Bi-monthly. A current-awareness service.

Industrial Aerodynamics Abstracts. British Hydromechanics Research Association, 1970–   . Bi-monthly. 900 informative abstracts p.a. Annual indexes are provided for retrospective searching.

Two other abstracting services are perhaps worth noting, although they cover a much wider field than fluid mechanics. The first is the fortnightly *International Aerospace Abstracts* (American Institute of Aero- and Astronautics, 1961–   ), which abstracts all relevant literature except reports; the second is *Scientific Technical and Aerospace Reports*, (National Aeronautics and Space Administration, 1963–   ), which covers report literature only. Both provide informative abstracts and are suitable for retrospective searching.

Transonic, supersonic and hypersonic flow is of theoretical interest because of its departure from classical hydrodynamics, i.e. incompressible flow occurs. The practical applications need no emphasising. For a comprehensive treatise on all aspects of high-speed flow there is the 12 volume work *High speed aerodynamics and jet propulsion* (OUP 1955–   ). Although concerned primarily with high-speed flight, its approach is very fundamental.

Transonic flow is adequately covered in *Transonic aerodynamics*, by C. Ferrari and F. Tricomi (Academic Press, 1968). Mainly concerned with the flow past airfoils.

Developments on hypersonic flow are dealt with in *Modern developments in gas dynamics*, by W. H. T. Lok (Plenum, 1969). For the experimentalist there is *Experimental methods of hypersonics*, by J. Lukasiewicz (Dekker, 1973).

Rarefied gas dynamics is clearly concerned with the flow of gases at low pressures, as in vacuum systems, in space, etc., less obvious perhaps is its applicability at normal pressures when the time scale of an event is comparable with the molecular mean time of flight, as in shock-wave and some ultrasonic phenomena. A recent book giving an up-to-date treatment, is *Rarefied gas dynamics*, by M. N. Kogan (Plenum, 1969).

There appears to be no specialist periodical literature; papers appear in the wider-based journals and are abstracted by the usual fluid mechanics services. Recent developments appear in *Rarefied Gas Dynamics*, Proceedings of the International Symposia, 1958– . Published as *Advances in Applied Mechanics, Supplement*, Academic Press, 1958– . The most recent is the 6th, MIT, 1968.

Plasma fluid dynamics or magnetohydrodynamics is concerned with the flow of conducting media, particularly in the presence of magnetic fields. Introductory texts are *Theoretical magneto-fluid-dynamics*, by H. Cabannes (Academic Press, 1970), and *An introduction to magneto-fluid-mechanics*, by V. C. Farraro and C. Plumpton (OUP, 2nd edn, 1966).

The specialist periodical literature includes:

Journal of Plasma Physics. CUP, 1964– . Bi-monthly.

Magnetohydrodynamics. Faraday Press, 1965–   (1966–). Bi-monthly. Cover-to-cover translation of *Magnitnaya Girodinamika*, 1965– .

Plasma Physics. Pergamon, 1965– . Monthly. Papers in English, French or German. Summaries in English.

The principal specialist abstracting journals are:

Plasma Physics Index. Zentralstelle für Atomkernenergie-Dokumenta-tion, (Frankfurt) 1960– . Monthly. Provides over 5000 references p.a. Reports are listed separately. Retrospective searching facilitated by annual indexes.

Science Research Abstracts. Part A. Superconductivity; Magneto-hydrodynamics and Plasmas; Theoretical Physics. Cambridge (Mass.) Scientific Abstracts, Inc., 1973– . 10 issues per volume. 16 000 informative abstracts arranged in subject order. Annual indexes make this a useful tool for retrospective searching.

Turbulence, is an irregular eddying motion characteristic of fluid at high Reynolds Numbers. As such it is responsible for much wasted energy in any context where a fluid is moved at useful speeds, as in water mains, or where an object is moved through a fluid, as with aircraft. On the other hand, the phenomenon is put to good practical use in mixing and heat transfer.

Two good introductory texts are *An introduction to turbulence and its measurement*, by P. Bradshaw (Pergamon, 1971), and *Turbulence phenomena. An introduction to the eddy transfer of momentum, mass and heat, particularly at interfaces*, by J. T. Davies (Academic Press, 1972. Glossary).

More advanced and comprehensive treatments are to be found in

Hydrodynamic and hydromagnetic stability, by S. Chandrasekhar (Clarendon, 1961), and *Statistical fluid mechanics. Mechanics of turbulence*, by A. S. Mouin and A. M. Yaglom (MIT, 1971); this latter contains a 50-page bibliography. A recent book is *Development of the theory of turbulence*, by D. C. Leslie (Clarendon, 1973).

The computer approach to turbulence is concisely presented in *Lectures in mathematical models of turbulence* (Academic Press, 1972). Problems of cavitation are discussed in *Cavitation*, by R. T. Knapp, J. W. Daly and F. G. Hammit (McGraw-Hill, 1970), a lucid and up-to-date account of a complex subject, covering theory, experiment and application.

Specialist works on constrained flow, i.e. flow through tubes, channels, orifices, etc., and including jets, include *Theory of jets in ideal fluids*, by M. I. Gurevich (Academic Press, 1965); *The theory of turbulent jets*, by G. N. Abramovich (MIT, 1963); *Dynamics of fluids in porous media*, by J. Bear (American Elsevier, 1972); *Flow through porous media*, by De Wiest (Academic Press, 1969); and *Separation of flow*, by P. K. Chang (Pergamon, 1970).

A regularly occurring symposium on flow in porous media is *International Symposium on the Fundamentals of Transport Phenomena in Porous Media*, 1969– . The most recent was held in Haifa, 1972, and sponsored by the International Association for Hydraulic Research. The subject is discussed from many viewpoints, e.g. hydrology, reservoir-engineering, soil mechanics, chemical engineering, etc.

Multiphase flow is of great practical interest, principally in chemical engineering, and is concerned with both gas-liquid and solid–fluid flow. Recent works include *Flowing gas–solid suspensions*, by R. G. Boothroyd (Chapman and Hall, 1971), on the hydrodynamics of high concentration powder gas mixtures; *One-dimensional two-phase flow*, by G. B. Wallis (McGraw-Hill, 1969); and *Fluid dynamics of multiphase systems*, by S. L. Soo (Blaisdell, 1967).

A new bi-monthly periodical covering all aspects of multiphase flow is *International Journal of Multiphase Flow* (Pergamon, 1973–). Recent developments in solid–liquid flow may be followed in the quarterly publication *Solid–liquid Flow Abstracts* (British Hydromechanics Research Association, 1968–); this presents 400 informative and indicative abstracts each year, culled from 200 journals. Annual indexes make it suitable for retrospective searching.

Flow measurement, experimental techniques and instrumentation are covered by the general literature of mechanics. However, specialist literature exists. Works of interest include *The theory of electromagnetic flow measurement*, by J. A. Shercliffe (CUP, 1962), and *The measurement of air flow* (Pergamon, 1966).

A useful bibliography covering the measurement of flow in open channels during the period 1950–70 is *Fluid flow measurement. A bibliography*, by R. R. Dowden (British Hydromechanics Research Association, 1972). It contains 2400 abstracts from all forms of literature on a world-wide basis; entries arranged by subject.

In view of the rapidly increasing use of the computer in all branches of science and its effects on the theoretical approach to the subject, access to proved programs is becoming more and more necessary. *A catalogue of digital computer programs in fluid mechanics* (Engineering Sciences Data Unit, London 1972) is a useful guide.

Shock waves are closely associated with supersonic, plasma and rarefied gas flow, and are frequently used as probes in many areas of science. The subject may be studied in *Stossröhre*, by H. Oertel (Springer, 1966), in German, an important work with a useful bibliography and glossary; *Shock waves and detonation in gases*, by R. I. Soloukhin (Mono Book, 1966); *Relaxation in shock waves*, by Ye. V. Stupochenko, Ye. V. Loser and A. I. Osipin (Springer, 1971); and *Physics of shock waves and high temperature physics*, by Ya. B. Zel'dovich and Yu. P. Raizer (Academic Press, 1966, 1967, 2 vols.).

More recent work is presented at regular intervals in *International Shock Tube Symposium*. The latest conference was the 8th, London, 1971 (Chapman and Hall, 1971).

Vacuum techniques

Vacuum techniques are extensively used, forming an important part of the experimentalist's and process engineer's repertoire. Applications which come to mind are the manufacture of electronic devices, the simulation of space conditions, plasma physics, and so on.

Among the general books covering the technology of low- and high-vacuum with or without ultra-high vacuum, may be listed:

Atta, C. M. van. *Vacuum science and engineering*. McGraw-Hill, 1965

Beck, A. H. W. *Handbook of vacuum physics*. Pergamon, 1964– . In three volumes.

Diels, K. and Jaeckel, R. *Leybold vacuum handbook*. Pergamon, 1966. Contains an extensive bibliography.

Dushman, S. *Scientific foundations of vacuum techniques*. 2nd edn, Wiley, 1962

Pirani, M. and Yarwood, J. *Principles of vacuum engineering*. Chapman and Hall, 1961

Yarwood, J. *High vacuum technique, theory and practice and properties of materials*. Chapman and Hall, 1967. An introductory text.

Among the works devoted solely to ultra-high vacuum mention may be made of *The physical basis of ultra-high vacuum*, by P. A. Redhead, J. P. Hobson and E. V. Kornelsen (Chapman and Hall, 1968), and *The physical principles of ultra-high vacuum*, by N. W. Robinson (Chapman and Hall, 1968).

Conferences have been held regularly since 1954 by the American Vacuum Society under the title *National Symposium on Vacuum Technology*. The proceedings of the most recent conference—the 16th, 1970—appear in the *Journal of Vacuum Science and Technology*, 1970, Vol. 7. Abstracts of papers presented at earlier conferences appear in the earlier issues of the same journal.

Another regular conference is the *International Vacuum Congress. Proceedings*, 1958– . The most recent was the 5th, Boston, 1971, published in *Journal of Vacuum Science and Technology*, 1972, Vol. 9, No. 1.

In addition to the comprehensive works listed above, there are many devoted to some particular aspect of the subject. Among those devoted to vacuum materials two should be mentioned in particular. Firstly, *Materials of high vacuum technology*, by W. Espe (Pergamon, 1966–), in three volumes covering metals and metalloids, silicates, and auxiliary materials, respectively; an important work with extensive references and considerable numerical data. Secondly, *Handbook of materials and techniques for vacuum devices*, by W. H. Kohl (Reinhold, 1967).

For vacuum pumps there is *High vacuum pumping equipment*, by B. D. Power (Chapman and Hall, 1966); and for vacuum seals *Vacuum sealing tubes*, by A. Roth (Pergamon, 1966), describing in considerable details, seals to transmit electricity, motion and materials into vacuum systems during operation.

There are few journals on vacuum technology. They all contain papers on original work, brief communications, news, etc.

Journal of Vacuum Science and Technology. American Vacuum Society, 1964– . Volumes 1–6 contain the abstracts of the proceedings of the 12th–15th National Symposia on Vacuum Technology. The full proceedings of later symposia are published in the later volumes.

Le Vide. Société Française des Ingénieurs et Techniciens du Vide, 1946– . Bi-monthly. Also issues as supplements the proceedings of Colloque International sur les Application des Techniques du Vide à L'Industrie des Semiconducteurs, Colloque International Vide et Froid, and a number of 'journées.'

Vacuum. Pergamon, 1951– . Monthly. There is an abstract section presenting several hundred informative abstracts each year. The entries are arranged in subject order; there are no cumulated indexes, which restricts the use to current awareness only.

Vakuum Technik (formerly *Glas-und Hochvakuum-Technik*). Deutsche Arbeitsgemeinschaft Vakuum, 1952– . Articles are in German with English and French summaries. Contains an abstract section; 190 informative abstracts p.a. in German. Annual indexes.

Vuoto. Associazione Italiana del Vuoto, 1968– . Quarterly. Articles in English or Italian, with summaries in English. Recent patents are listed in each issue.

In addition to these journals, many others publish papers on vacuum science. Most of these are abstracted and indexed by the major abstracting services; to these may be added the abstract sections of *Vacuum* and *Vakuum Technik* mentioned earlier, and one 'specialist abstracting periodical—*Surface and Vacuum Physics Index*, Institut für Plasmaphysik, Zentralstelle für Atomkernenergie-Dokumentation, 1965– . A monthly indexing service with about 3000 entries p.a. Annual indexes facilitate retrospective searching.

The terminology of vacuum technology is covered by *Glossary of terms used in vacuum technology* (BSI 2951; British Standards Institute, 1958). An American glossary, with the same title, was also published in 1958 by Pergamon for the Committee on Standards of the American Vacuum Society. Two useful translation aids are:

Hurrle, K., Jablowski, F. M. and Roth, J. (eds.). *Technical dictionary of vacuum physics and vacuum technology*. Pergamon, 1972. Terms are arranged in English order, with French, German and Russian equivalents.

Weber, F. (ed.). *Elsevier's dictionary of high vacuum science and technology*. Elsevier, 1968. German terms arranged alphabetically, with English, French, Spanish, Italian and Russian equivalents.

Among the organisations concerned with vacuum science may be listed the International Union for Vacuum Science, Technique and Applications; the American Vacuum Society; and the Associazione Italiana del Vuoto.

SOUND: ACOUSTICS

Sound is a mechanical disturbance in an elastic medium. The term is usually applied to disturbances operating over a range of frequencies which produce an aural sensation. Below these frequencies the phenomenon is referred to infrasound; above, it is called ultra-sonics.

Since the disturbance is propagated through a medium as a wave, it exhibits the usual wave phemonena of absorption, refraction, diffraction, reflection, etc., and these are used to study sound itself, to investigate other physical phenomena or to form the basis of instrumentation in some branch of technology, e.g. non-destructive testing, sonar, etc.

Upon examination the literature of acoustics is found to fall into a number of distinct sections, i.e. general acoustics, architectural acoustics, noise, audio, music, ultrasonics, underwater acoustics, physiological acoustics, speech, bio-acoustics, and so on. Most of these headings are used here.

A good general coverage of acoustics is to be found in the following works.

Encyclopedia of physics. Vol XI/1 and 2, Acoustics (see p. 37). A series of extensive articles on modern aspects of acoustics.

Morse, P. N. *Vibration and sound.* McGraw-Hill, 1936

Stephens, R. W. B. and Bate, A. E. *Acoustics and vibrational physics.* 2nd edn. Arnold, 1966. An advanced graduate text.

Olson, H. F. *Acoustical engineering.* van Nostrand, 1957. Treats the fundamentals of applied acoustics.

Richardson E. G. and Meyer E. (eds.). *Technical aspects of sound.* Elsevier, 1953–62. 3 vols. Contains a number of papers by acknowledged experts covering a wide range of topics in applied acoustics. Volume 3 updates the previous two volumes.

A recent enterprise is the *Benchmark Papers In Acoustics*, a series of monographs published by Dowden, Hutchinson and Ross Inc., Each volume is devoted to some particular aspect of acoustics and comprises a collection of seminally important papers, carefully selected and arranged by experts. The volumes summarise the existing state of knowledge and also act as a foundation for further work. The three volumes so far published are *Underwater acoustics,* edited by V. M. Albers; *Acoustics. Historical and philosophical developments,* edited by R. B. Lindsay; and *Speech synthesis,* edited by J. L. Flanagan and L. R. Rabiner.

Recent developments are reviewed in the later volumes of *Physical acoustics. Principles and methods,* by W. P. Mason and R. N. Thurston (Academic Press, 1966–   . A multivolume work presenting a number of articles by different experts on the whole range of acoustic phenomena. Volume 9, 1972, is the most recent.

Some aspects of propagation are of sufficient interest to warrant separate treatment. One such aspect is wave guides, a theoretical treatment of which appears in *Mechanical waveguides. The propagation of acoustic and ultrasonic waves in fluids and solids with boundaries,* by M. Redwood (Pergamon, 1960). A recent general work on elastic waves is *Wave propagation in elastic solids*, by J. D. Achenbach (North-Holland, 1973). Scattering phenomena are treated in *Electromagnetic and acoustic scattering by simple shapes*, by J. J. Bowman and others (North-Holland, 1969).

Current interest in non-linear elastic waves is illustrated in *Wave propagation in solids*, Proceedings of the ASME Winter Annual Meeting, Los Angeles, 1969. Edited by J. Miklowitz and published by the American Society of Mechanical Engineers, 1969. Works treating the propagation of elastic waves through layered, turbulent, etc. media are listed under 'Oscillations, vibrations and waves'.

Measurement techniques may be studied in *Vibration and acoustic measurement handbook*, by M. P. Blake and W. S. Mitchell (Spartan Books, 1972).

Below are listed some of the more important general journals in acoustics. Each one covers some or all of the different topics which go to make up the subject. The more specialist literature is presented later.

Acoustica. A. Hirzel Verlag, 1951– . Monthly. Articles in English, French and German, with summaries in all three languages.

Journal of the Acoustical Society of America. The Society, 1929– . Monthly. Decennial indexes available. Contains a current awareness service. Entries are arranged in 17 sections. Reviews, bibliographies, patents and standards are presented together in one of the sections.

Journal of Sound and Vibration. Academic Press, 1964– . Fortnightly. Contains original papers, reviews and brief reports.

Proceedings of the British Acoustical Society. 1971– . Two issues p.a.

Revue d' acoustique. Groupement des acousticiens de langue française. 1932–40; 1968– . Quarterly. Contains summaries of the papers in English.

Soviet Physics. Acoustics. American Institute of Physics, 1955– . Quarterly. (Cover-to-cover translation of *Akusticheskii Zhurnal*.) Contains some review articles.

Two important Oriental journals which contain some English in the form of lists of contents or summaries are *Acta Acoustica Sinica* and *Journal of the Acoustical Society of Japan*.

In addition to these more formal publications, many institutions publish reports, memoranda, etc. Examples of such organisations are the Institute of Sound and Vibration, University of Southampton; Istituto di Acustica 'O. M. Corbino'; and British Ship Research Association.

Important developments are reported in the proceedings of the

International Congress on Acoustics. Held at three-yearly intervals, the most recent one was the 7th, Budapest, 1971, published by Akademiai Kiado, Budapest, 1971.

Other conferences and meetings are held by numerous societies and institutions throughout the world. These societies may also be approached for information and advice. Examples of such organisations are the Acoustical Society of America, the British Acoustical Society and the Naval Undersea Center-San Diego.

Acoustics is well covered by the general abstracting services for physics. There is also *Acoustics Abstracts* (Multi-Sciences Publishing Company, 1967–   ) a bi-monthly specialist service presenting about 1200 brief entries per annum with annual indexes. In addition there is a regular indexing service in the *Journal of the Acoustical Society of America.*

A recent list of definitions of 500 acoustical terms is given in *Glossary of acoustical terms* (BSI 661: 1969; British Standards Institute, 1969). Of particular interest to acousticians is *Elsevier's dictionary of cinema, sound and music*, edited by W. E. Clason (Elsevier, 1956). Over 3000 entries are arranged in English order, with equivalents in Dutch, French, Spanish, Italian and German. There is also a useful bibliography of dictionaries, glossaries, terminologies, nomenclatures and textbooks. For Russian there is *Russian-English glossary of acoustics and ultrasonics*, by P. Robeson (Consultants Bureau, 1958). Terms listed in Russian order with English equivalents. There is also a list of Russian equivalents of names found in acoustics theory.

Architectural acoustics

Architectural acoustics is concerned with the satisfactory propagation of intelligible sound within a confined space. It has close associations with Noise, which is dealt with below. Recent texts are *Acoustic design and practice*, by R. L. Suri (Asia Publishing House, 1966)—particularly useful for its glossary and numerical data; and *Sound structures and their interactions*, by M. Junger and D. Feit (MIT, 1972, Glossary).

Current work is indexed by *RIBA Annual Review of Periodical Articles* in addition to the usual acoustics abstracting services.

Noise

Noise is unwanted sound. In buildings it can arise from echoes and reverberations and can to some extent be controlled by design. In other environments it arises from vibrations caused by machinery, aerodynamic effects, and so on. It has long been recognised that noise is a nuisance detrimental to health, and considerable efforts are made to control it—by legislation and by physical means. The words 'buildings', 'machinery', 'aerodynamic', 'environment', 'health' and 'legislation' used above give some idea of the range of subjects to be explored, if all aspects are to be covered. Here we are primarily concerned with the physical aspects—namely origin, transmission, measurement and control.

Among the books which treat the subject comprehensively may be mentioned *Noise and vibration control*, by L. L. Beranek (McGraw-Hill, 1971); *Handbook of noise control*, edited by C. M. Harris (McGraw-Hill, 1957); and *Guidelines on noise* (Tracer House Inc., 1973, for the American Petroleum Institute)—a short informative guide with a useful section on reference data and standards. For those concerned with the design of noise-free machinery there is *Acoustics design and noise control* (Chemical Publishing, 1973). Finally a trade directory which also contains a useful collection of physical data—*Handbook of noise vibration control* (2nd edn, Trade and Technical Press, 1973).

The measurement of noise is covered in the general literature of the subject and also by specialist books such as *Handbook of noise measurement*, by A. P. G. Peterson and E. E. Gross (4th edn., General Radio Company, 1960); *Noise measurement techniques* (National Physical Laboratory, HMSO, 1955); and *Vibration and acoustic measurement*, by M. P. Blake and W. S. Mitchell (Spartan Books, 1972).

Among the specialist journals, including the proceedings of regularly held conferences, are:

Applied Acoustics. Elsevier, 1968– . Quarterly. Intended for technologists, architects, public health inspectors, etc., responsible for the design and construction of industrial equipment, buildings and transport systems.

Kampf dem Lärm. Deutscher Arbeitsring für Lärmbekämpfung, 1954– . Bi-monthly.

Noise Control and Vibration Reduction. Trade and Technical Press, 1970– . Quarterly. Intended for designers of machines and buildings. Contains a bibliography, buyer's guide and directory of sources of specialised assistance.

Recent conferences have been held by the University of Wales Institute of Science and Technology (1973), Purdue University (1972), University of Toronto Institute for Aerospace Studies (1969),

International Association against Noise (Groningen, 1960) and Institute of Noise and Control Engineering (Washington 1972).

Abstracting services specialising in the literature of noise are:

Noise and Vibration Bulletin. Multi-Science Publishing, 1970– . Monthly. 400 informative abstracts appear each year, selected from 1000 journals. A short annual index of titles makes this a possible though difficult tool for retrospective searching.

Noise Bibliography. Technology Reports Centre. Ministry of Technology 1962– . Annual. Abstracts are informative, mainly of reports, or papers given report status, and mainly concerned with aircraft noise. There are no cumulative indexes, which restricts its use to current awareness.

Shock and Vibration Digest. Naval Research Laboratory, Shock and Vibration Information Centre, Washington, 1969– . Monthly. About 1800 informative abstracts each year arranged in subject classified order, with annual indexes. There is also a section containing longer critical reviews of items abstracted earlier.

Transportation Noise Bulletin. Transportation Noise Research Information Service, US National Academy of Science, 1972– . Comprises abstracts of report literature.

Ultrasonics

Ultrasonics is that branch of acoustics involving frequencies greater than about 20 kHz. Currently the upper limit of ultrasonic frequencies is in the gigaherz region, corresponding to the microwave region of the electromagnetic spectrum. With this range of frequencies ultrasonic radiation exhibits a wide range of phenomena and finds a large and increasing number of scientific and technological applications, such as molecular, physics, communications, sonar, non-destructive testing, process control, medical diagnostics.

The following books will provide a comprehensive introduction:

Beyer, R. T. and Letcher, S. V. *Physical ultrasonics.* Academic Press, 1969. An advanced introductory text.

Ensminger, D. *Ultrasonics. The low and high intensity applications.* Dekker, 1973. Reviews the literature of recent developments in applied ultrasonics. Glossary.

Gooberman, G. L. *Ultrasonics. Theory and application.* English Universities, 1968. A short introductory text.

A detailed view of recent developments is available in the monograph series *Ultrasonic technology,* edited by L. Balamuth (Plenum, 1967–). The four volumes available so far are *Rayleigh and Lamb waves* (I. A. Viktorov, 1967). *Sources of high intensity sound* (L. D. Rozenberg, 1969, 2 vols), *Ultrasonic transducer materials* (O. E. Mattiat, 1971) and *High intensity ultrasonic fields* (L. D. Rozenberg, 1971). Each volume contains several authoritative articles on specific topics.

Technical applications are comprehensively treated in *Ultrasonic engineering*, by J. R. Frederick (Wiley, 1965).

The specialist periodical literature includes the following titles:

IEEE Transactions on Sonics and Ultrasonics. Institute for Electrical Engineering and Electronics, 1964– . Quarterly. Emphasises the application of fundamental knowledge. Contains reviews and data for design practice.

Russian Ultrasonics. Multi-science Publishing, 1971– . Quarterly. Contains original translations from a large number of Russian journals, many not widely known in the West. There are about 60 papers per issue, covering all aspects of ultrasonics.

Ultrasonics. IPC Science and Technology Press, 1963–. Quarterly. Also contains two to three pages of brief informative abstracts of patents.

Recent conferences include *International Symposium on High Power Ultrasonics* (1st, Graz, 1970), *Ultrasonics Symposium* (New York, 1970) and *International Conference on Non-destructive Testing* (6th, Hanover, 1970).

The literature of ultrasonics is abstracted by the general services for physics and acoustics. In addition there is *Ultraschall-Dokumentation*, Laboratorium für Ultraschall, Rheinisch-Westfälische Technische Hochschule, Aachen, 1965(?), irregular, every 2 or 3 years. The 7000 indicative entries in each volume are arranged in classified subject order. Special forms of literature such as reviews, bibliographies, dictionaries, conference proceedings, etc., are clearly indicated.

Molecular acoustics is concerned with the interaction of ultrasonic radiation with molecules. Recent books covering the subject in some depth include *Absorption of ultrasonic waves*, by W. F. Herzfeld and T. A. Litovitz (Academic Press, 1959); *Molecular acoustics*, by A. J. Matheson (Wiley-Interscience, 1971); and *Ultrasonic absorption. An introduction to the theory of sound absorption and dispersion in gases, liquids and solids*, by A. B. Bhatia (Clarendon, 1967).

An important source of numerical data is *Landolt-Börnstein numerical data and functional relationships in science and technology. New series. Group 2. Atomic and molecular physics. Volume 5. Molecular acoustics,* (see p. 43). This work tabulates data for the velocity and absorption of ultrasound in gases, liquids and solids, and velocity of shock waves in materials.

Non-destructive testing (NDT) is probably the most important industrial application of ultrasonics. A recent treatment is given in *Ultrasonic testing of materials*, by J. and H. Krautkrämer (Springer, 1969). Recent developments are reported in the periodical literature of acoustics and also in the non-destructive testing journals, such as *The British Journal of Non-Destructive Testing, Soviet Journal of Non-destructive Testing* (cover-to-cover translation of *Defektoskopiya*), *Materials Evaluation, Non-Destructive Testing* and *International Journal of Non-destructive Testing*.

Review articles appear in *Progress in Applied Materials Research* (Heywood, 1900–).

The literature is abstracted in the usual services for physics and acoustics and also in the engineering abstracting services such as *Engineering Index Monthly*.

The terminology of ultrasonic non-destructive testing together with foreign-language equivalents is to be found in *List of terms used in ultrasonic testing* (Institute of Welding, 1967). Terms are arranged in subject order and defined in English and French. The foreign-language equivalents are given in Danish, Dutch, German, Italian, Polish, Slovenian, Spanish, Swedish and Turkish.

Surface wave phenomena have long been known, but interest has recently grown—particularly in surface wave devices operating at gigaherz frequencies. In addition to I. A. Viktorov's book mentioned earlier, there are: *Acoustic surface wave and acousto-optic devices*, edited by T. Kallard (Optosonic Press, 1971), a state of the art review; and *Microwave ultrasonics in solid state physics*, by J. W. Tucker and V. W. Rampton (North-Holland, 1972).

At present current papers are published in the periodical literature of acoustics and ultrasonics, and abstracted in the usual abstracting services. In addition there is *Surface Wave Abstracts*, Multi-Science Publishers, 1971– . Quarterly. 160 informative abstracts are culled from the world's literature each year. The entries are arranged by subject—one of them concerned with reviews and bibliographies—and there are annual indexes.

Underwater acoustics is a rapidly growing field of activity concerned with the propagation of both audio and ultrasonic radiation. It is of interest as a phenomenon in its own right and also because of its applications to many branches of science and technology, such as oceanology, underwater communication, submarine detection, etc.

Advanced introductory texts include *Ocean acoustics. Theory and experiment in underwater sound*, by I. Tolstoy and C. S. Clay (McGraw-Hill, 1966); *Physics of sound in the sea* (Gordon and Breach,

1968, 4 vols.); *Underwater acoustics handbook*, by V. M. Albers (Pennsylvania University Press, 1965); *Underwater acoustics*, edited by V. M. Albers (Benchmark Papers in Acoustics, Dowden Hutchinson and Ross, 1972); and *Underwater acoustics*, by L. Camp (Wiley-Interscience, 1970). A recent collection of papers surveying the field appears in *Underwater acoustics*, by R. W. B. Stephens (Wiley-Interscience, 1970).

Details of underwater measurements are to be found in *Underwater electroacoustic measurements*, by R. J. Bobber (Naval Research Laboratory, Washington, 1970).

There is no specialist periodical literature as yet. Current work is reported chiefly in the acoustics literature or in the proceedings of conferences. Two such are:

Institute on Underwater Acoustics, which has met twice—in London, 1961 and Copenhagen, 1967; the Proceedings were published by Pergamon, 1963, 1967.

Marine Bioacoustics. Symposia held in Bimini 1963, New York, 1966.

Other papers appear in the literature of oceanography, such as *Ocean Engineering, Journal of Physical Oceanography*. At present the physics and acoustics abstracting services provide the only reasonable access to current papers.

11

Heat and thermodynamics

J. W. Burchell

GENERAL BIBLIOGRAPHIES AND ABSTRACTS JOURNALS

The classification adopted by *Physics Abstracts* (*Science Abstracts*, *Series A*) spreads the various aspects of heat and thermodynamics through several subject classes. In addition to the section on 'Statistical Physics and Thermodynamics', abstracts on thermal instruments and techniques, including those for high and low temperatures, are found under 'Instrumentation and Experimental Techniques', and the thermal properties of matter are included in the classes 'Physics of Gases' and 'Condensed Matter: Structure, Thermal and Mechanical Properties'. The French abstracts journal *Bulletin Signalétique* puts references to heat and thermodynamics in *Section 130: Physique*. *Engineering Index* is mainly concerned with instrumentation and applied aspects of the subject, such as heat transfer, but it does include some abstracts on theoretical thermodynamics. A bibliographical serial devoted entirely to heat and thermodynamics is the annual *Heat Bibliography* prepared by the National Engineering Laboratory and published by Her Majesty's Stationery Office. It contains references noted in the laboratory during each year; the majority are from abstracting and bibliographical journals, and in such cases the source is given with an indication of whether or not an informative abstract is included in the original journal. References are separated into broad subject divisions such as 'Applications and Plant Equipment', 'Chemical and Physical References to Particular Substances', 'Energy Conversion and Thermodynamics of Cells', 'Physical and Thermal Properties', etc. The bias is naturally towards engineering, but it does contain references to physical and theoretical work as well.

TEXTBOOKS, MONOGRAPHS AND CONFERENCES

Two popular textbooks at an undergraduate level which have run to several editions are *Heat and thermodynamics*, by J. K. Roberts and A. R. Miller (5th edn, Blackie, 1960) and *Heat and thermodynamics: an intermediate textbook*, by M. W. Zemansky (5th edn, McGraw-Hill, 1968). Several other standard texts on thermodynamics exist, and although some of them were published a number of years ago they still retain their value. Of these the following, written mostly for the advanced undergraduate/graduate student, are to be recommended:

Guggenheim, E. A. *Thermodynamics: an advanced treatment for chemists and physicists*. 5th revised edition. North-Holland, 1967.

Landsberg, P. T. *Thermodynamics with quantum statistical illustrations*. Interscience, 1961.

Pippard, A. B. *Elements of classical thermodynamics for advanced students of physics*. CUP 1957.

Sommerfeld, A. *Thermodynamics and statistical mechanics*. Academic Press, 1956.

Wilson, A. H. *Thermodynamics and statistical thermodynamics*. CUP, 1957.

A further book concerned only with non-statistical thermodynamics which is intended to supplement these standard texts is *The concepts of classical thermodynamics*, by H. A. Buchdahl (CUP, 1966), and a slightly more recent text which discusses both classical thermodynamics and equilibrium statistical mechanics is L. M. Grossman's *Thermodynamics and statistical mechanics* (McGraw-Hill, 1969). The fundamental postulates of thermodynamics have a very general application in science, as is demonstrated in *Principles of general thermodynamics*, by G. N. Hatsopoulos and J. H. Keenan (Wiley, 1965), which includes a revision of the definitions and laws of classical thermodynamics, and reconciles the bases of thermodynamics and statistical mechanics. Also at the theoretical level is a book written for the graduate physicist, mathematician or philosopher—*Mathematical foundations of thermodynamics*, by R. Giles (Pergamon, 1964).

As will be seen from their titles, several of the books mentioned above discuss both classical (macroscopic) and statistical (microscopic) thermodynamics. The following books concentrate on non-classical thermodynamics and its applications:

Haase, R. *Thermodynamics of irreversible processes*. Addison-Wesley, 1969. Written partly for physical chemists, it includes the application of theory to chemical reactions, as well as to electromagnetic fields and anisotropic materials.

Kelly, D. C. *Thermodynamics and statistical physics: an elementary treatment with contemporary applications.* Academic Press, 1973

Münster, A. *Statistical thermodynamics.* 1st English edition. Springer/ Academic Press, Vol. 1, 1969. A revised and enlarged version of the 1956 German edition, which surveys the entire field and emphasises recent developments.

Sonntag, R. E. and Van Wylen, G. J. *Fundamentals of statistical thermodynamics.* 2nd edition. Wiley, 1973. Intended as a textbook for engineering students, it should also be useful for physicists.

Yourgrau, W. *et al. Treatise on irreversible and statistical thermophysics: an introduction to nonclassical thermodynamics.* Macmillan, 1966. For the advanced undergraduate or graduate student, it contains both theory and applications.

The published proceedings of two recent conferences provide reviews of current developments in various aspects of thermodynamics. The first of these, *Proceedings of the International Symposium: Critical Reviews of the Foundations of Relativistic and Classical Thermodynamics, April 7–8, 1969, Pittsburgh*, edited by E. B. Stuart *et al.* (Mono Book, 1970), covers such topics as deduction quantum thermodynamics, the third law and statistical mechanics, and relativistic statistical thermodynamics. The proceedings of the second, *International conference on thermodynamics, Cardiff 1970*, were published in *Pure and Applied Chemistry*, Vol. 22, pp. 215–570 (1970). The subjects considered at that meeting included the foundations of thermodynamics, thermomechanics, irreversibility and quantum mechanics, and statistical thermodynamics in astrophysics and relativity.

One of the most important concepts which arises in thermodynamics is that of entropy, which is discussed in detail in the monograph *Entropy: the significance of the concept of entropy and its applications in science and technology*, by J. D. Fast (2nd edn, N. V. Philips, Eindhoven, 1968). The methods of classical statistical mechanics used in the study of entropy take up the major part of J. Yvon's *Correlations and entropy in classical statistical mechanics* (Pergamon, 1969). The third law of thermodynamics is concerned with the behaviour of entropy and internal energy as the temperature approaches absolute zero. This is the subject of *The third law of thermodynamics*, by J. Wilks (OUP, 1961) and *Entropy and low temperature physics*, by J. S. Dugdale (Hutchinson, 1966).

THE MEASUREMENT OF HEAT—THERMOMETRY AND CALORIMETRY

The international committee which is concerned with thermometry and the standardisation of temperature scales is the *Comité Consultatif de Thermométrie* which comes under the *Comité International des Poids et Mesures*. The proceedings [*Travaux*] of the Sessions of the *Comité Consultatif* are published, mainly in French. Sessions are held every three or four years, the most recent being the ninth, held in 1971. This committee is responsible, among other things, for the International Practical Temperature Scale (IPTS) which provides the practical realisation of thermodynamic temperatures, and is the scale on which standardising laboratories and others should base their measurements. The official French text for the latest (1968) scale was published as Annexe 18 to the proceedings of the 8th Session, 1967, of the *Comité Consultatif de Thermométrie* and also as Annexe 2 to *Comptes Rendus des Séances de la Treizième Conférence Général des Poids et Measures*, 1967. The text in English was published in the journal *Metrologia*, Vol. 5, No. 2, pp. 35–44 (1969), and also by the National Physical Laboratory as a booklet entitled *The International Practical Temperature Scale of 1968* (HMSO, 1969). The same Laboratory has produced another booklet—*The Calibration of Thermometers*, by C. R. Barber (HMSO, 1971)—which summarises the way in which the IPTS is defined and gives details of how to calibrate various types of thermometer at fixed points or by comparison with previously calibrated instruments. It includes a short, but useful, bibliography.

The journal just mentioned, *Metrologia*, is published under the auspices of the *Comité International des Poids et Mesures*, and contains the results of research into the fundamental measurements in all fields of physics, including thermometry. The infrequent symposium series *Temperature, its Measurement and Control in Science and Industry* contains papers on temperature scales, instrumentation, temperature control, and methods of temperature measurement in various sciences and technologies. The most recent volume (Vol. 4) contains the proceedings of the fifth Symposium on Temperature, Washington, 1971, and was published by the Instrument Society of America in 1972. Volumes 1 to 3 were published by Reinhold in 1941, 1955 and 1962, respectively, and contain the proceedings of the second, third and fourth Symposia. (The proceedings of the first in this series of symposia—Symposium on Pyrometry, New York City, 1919—were not published collectively.)

Of the various books on the subject *Instrumentation: temperature,* by F. E. Doyle and G. T. Byrom (Blackie, 1970), is a short introduction for technical students to the different types of thermometer and methods of calibration and testing; *Fundamentals of temperature, pressure and flow measurements*, by R. P. Benedict (Wiley, 1969), is intended for engineers and covers the practical methods of thermometry in some detail; *Thermocouple temperature measurement*, by P. A. Kinzie (Wiley, 1973), is a comprehensive summary of data and literature references for about 300 types of thermocouple. Other books include *Temperature measurement*, by J. A. Hall (Chapman and Hall, 1967), and *Technische temperaturmessungen*, by H. Lindorf (4th edn, Girardet, Essen, 1970). The National Bureau of Standards is responsible for all aspects of standardisation and measurement in the USA and has consequently published many papers on thermometry. *Precision measurement and calibration: selected NBS papers on temperature* (US Department of Commerce, NBS Special Publication 300, Vol. 2, 1968) contains reprints of papers by NBS staff on temperature scales and various methods of thermometry.

One of the related technical problems is that of temperature control. *Thermostatic control: principles and practice*, by V. C. Miles (Newnes, 1965), covers the theory, operation and limitations of several devices. Similar coverage is found in *Fundamentals of temperature control*, by W. K. Roots (Academic Press, 1969), and problems which arise in special environments, such as spacecraft, are discussed in *Temperature control*, by M. Kutz (Wiley, 1968).

In addition to the publication on temperature mentioned above, the National Bureau of Standards has also produced a volume of previously published papers written by its staff on calorimetry. Edited by D. C. Ginnings, the title is *Precision measurement and calibration: selected NBS papers on heat* (US Department of Commerce, NBS Special Publication 300, Vol. 6, 1970) and it covers general calorimetry and techniques, calorimetry at high and low temperatures, heat transfer, etc. The Société Française des Thermiciens has published the book *Études sur la calorimétrie* (La Société, 1971), which describes the methods of calorimetry and their use in determining the thermodynamic properties of substances. *Recent progress in microcalorimetry*, by E. Calvet and H. Prat (Pergamon, 1963), describes the theory, use and applications of the Tian–Calvet microcalorimeter, which can be used to measure minute heat outputs over a long period of time and can be adapted to operate either at low temperatures or up to 1000 °C. The same subject is covered by the conference *Développements Récents de la Microcalorimétrie et de la Thermogenèse. Actes de la Colloque, Marseille 1965* (CNRS, Colloques Internationaux N. 156, 1967). Another conference, published as *Proceedings of the First International Conference on Calorimetry and*

Thermodynamics, Warsaw, 1969 (Polish Scientific Publishers, 1970) was chiefly concerned with chemical thermodynamics, but papers on techniques and apparatus were included.

THE TRANSFER OF HEAT—CONDUCTION, CONVECTION AND RADIATION

Most of the literature on heat transfer has been written by and for engineers who are concerned with the practical problems, but, as observed elsewhere, the difference between engineering (especially 'theoretical engineering') and physics is not always easy to define. The periodicals and books which follow should contain material of interest to anyone concerned with the physics of these processes.

There are two monograph series concerned with the transfer of heat. *Advances in Heat Transfer* (Academic Press) began in 1954, is published annually, and contains review articles or monographs on special topics of current interest. The other one, *Progress in Heat and Mass Transfer* (Pergamon), began in 1969 and has so far produced two volumes a year; particular mention should be made of Volume 3, 1971, which is titled *Heat transfer reviews 1953–69* and contains a review and list of the literature published in each of those years. The monthly *International Journal of Heat and Mass Transfer* (Pergamon) publishes both theoretical and experimental papers, and contains regularly an international bibliography of recently published papers. The *Proceedings of the Heat Transfer and Fluid Mechanics Institute* which is printed and distributed by Stanford University Press, reports on meetings of the Institute held annually until 1968 and thereafter biennially. The periodical *Wärme—und Stoffübertragung* (Springer) contains research papers mostly in German, and occasionally in English. Two journals which consist of translations of Russian and Japanese articles, respectively, not published elsewhere in English are *Heat Transfer—Soviet Research* and *Heat Transfer—Japanese Research* (Scripta).

Several of the books on heat transfer are translations of original Russian texts, including *A concise encyclopedia of heat transfer*, by S. S. Kutateladze and V. M. Borishanskii (Pergamon, 1966), and *Theory of heat and mass transfer*, by A. V. Lykov and Yu. A. Mikhailov (Israel Program for Scientific Translations, 1965). The last of these uses irreversible thermodynamics to develop an analytical theory of heat and mass transfer in gaseous mixtures, dispersed systems and porous bodies. Another book which is concerned with the fundamental theory is *Heat transfer*, by A. J. Chapman (2nd edn,

Macmillan, 1967); intended for engineering courses, it considers the three basic means of heat transfer separately and in various combinations.

The following books concentrate on one or other of the basic transfer processes.

Conduction

Proceedings of the International Conference on Thermal Conductivity. Held annually and concerned with all aspects of the subject—high and low temperatures, methods of measurement, various materials, etc.

Carslaw H. S. and Jaeger J. C. *Conduction of heat in solids.* 2nd edition. OUP, 1959. A classic work on the mathematical theory.

Tsederberg, N. V. *Thermal conductivity of gases and liquids.* Arnold, 1975. Explains the techniques and theories underlying experimental methods of measuring thermal conductivities and includes a considerable amount of data.

Tye, R. P. *Thermal conductivity.* 2 vols. Academic Press, 1969. A critical review of the methods for measuring thermal conductivity. Covers theoretical and practical details, all physical states of matter and the entire existing temperature range.

Convection

Borishanskii, V. M. and Paleev, I. I. (eds.). *Convective heat transfer in two-phase and one-phase flows.* Israel Program for Scientific Translations, 1969. A collection of theoretical and experimental papers for those doing research and development work on heat exchangers, steam generators, nuclear reactors and various related devices.

Institution of Mechanical Engineers. *Heat and Mass Transfer by Combined Forced and Natural Convection.* The Institution, 1972. Proceedings of a symposium held on 15 September, 1971.

Radiation

Hammond, H. K. and Masson, H. L. (eds.). *Precision measurement and calibration: selected NBS papers on radiometry and photometry.* US Department of Commerce, NBS Special Publication No. 300,

Vol. 7, 1971. A collection of over 60 papers published between 1957 and 1970 on general radiometry, emissivity standards, measurements and techniques, flux measurements, etc.

Hottel, H. C. and Sarofim, A. F. *Radiative transfer*. McGraw-Hill, 1967. A text and reference book giving principles, data on surface and gas radiation, techniques for calculating geometrically complex enclosures, and the application of principles to a variety of problems.

Özisik, M. N. *Radiative transfer and interactions with conduction and convection*. Wiley/Interscience, 1973.

Siegel, R. and Howell, J. R. *Thermal radiation heat transfer*. McGraw-Hill, 1972. A fairly comprehensive book divided into three sections— radiative behaviour of materials, radiation between surfaces and gas radiation.

THERMAL STRESSES

The effect of temperature differences on structures and materials is of considerable importance in many engineering situations. The following monographs have been selected for the basic physical theory which they include:

Boley, B. A. and Weiner, J. H. *Theory of thermal stresses*. Wiley, 1960. Concerned with the effects of large temperature differences, it discusses thermoelasticity, heat transfer theory and the theory of temperature effects in inelasticity.

Boley, B. A. (ed.). *Thermoinelasticity—Symposium of the International Union of Theoretical and Applied Mechanics, June 25–28, 1968*. Springer, 1970. Presents a comprehensive view of the behaviour of solids and structures at high temperatures under all types of inelastic regimes.

Johns, D. J. *Thermal stress analysis*. Pergamon Press, 1965. An introductory text mainly restricted to large variations in temperature and materials whose properties are time-independent.

Kovalenko, A. D. *Thermoelasticity: basic theory and applications*. Wolters-Noordhoff, Groningen, 1969. Particularly concerned with the laws and methods used in the study of thermal stresses in structural components produced by steady and unsteady temperature fields.

Yates, B. *Thermal expansion*. Plenum, 1972. Summarises recent advances including developments in lattice dynamics; experiments for measuring small displacements, especially at low temperatures; and optical, X-ray and electrical methods for measuring thermal expansion.

HIGH-TEMPERATURE PHYSICS

There are two bibliographical serials which must be mentioned, although neither is limited to just the physics of high temperatures. *Bibliography on the High Temperature Chemistry and Physics of Materials* is published by the International Union of Pure and Applied Chemistry, Commission on High Temperatures and Refractory Materials; it contains references only, not abstracts, and is not indexed. *High Temperature Bulletin* is published by the Information Centre on High Temperature Processes, Department of Fuel Science, The University, Leeds. It contains literature references and brief abstracts on a wide variety of topics including lasers, properties of materials, plasma physics, radiation and heat transfer, temperature measurement, etc.

Several journals are devoted to the study of high temperatures. *High Temperature Science* (Academic Press) includes papers on all aspects of the subject—chemistry, metallurgy, physics and engineering, including the production, measurement and control of high temperatures; thermodynamics; devices; and so on. *High Temperatures—High Pressures* (Pion, London) has a similar coverage and papers are published in English, French or German. The *Proceedings of the International Symposium on High Temperature Technology* are biased towards chemical aspects of the subject but research on thermodynamics and physical properties at high temperatures is included. The first symposium was held in 1959 and published by McGraw-Hill in 1960; the second and third were held in 1963 and 1967 and were published by Butterworths in 1964 and 1969, respectively. Volumes of *Progress in high temperature physics and chemistry* are published every year or two, and each contains a few lengthy papers on topics of current interest. The French journal, *Revue Internationale des Hautes Températures et des Réfractaires* (Masson), concentrates mainly on the physical and chemical properties of materials at high temperatures. Russian work is reported in *Teplofizika Vysokikh Temperatur*, which is translated cover-to-cover under the title *High Temperature* (1963– . Plenum); it includes plasma research, thermophysical properties of substances; heat transfer; heat and temperature measurement and high-temperature apparatus and its design.

LOW-TEMPERATURE PHYSICS

Cryogenics has attracted considerable attention in recent years and this is reflected in the number of periodicals and books devoted to the subject. To some extent this is due to interest in superconductivity and its practical possibilities, the literature of which has been considered in the section on electricity and magnetism.

The abstracts journal *Bulletin of the International Institute of Refrigeration* (also entitled *Bulletin de l'Institut International de Froid* (Paris) since it is published in both English and French) covers all aspects of refrigeration, including very low temperatures. The Institute did also produce the *Bibliographic Guide to Refrigeration*. Three volumes were published for the Institute by Pergamon Press covering the periods 1953 to 1960, 1961 to 1964, and 1965 to 1968, and consisting of bibliographical references to all documents abstracted in the *Bulletin* during those periods. From 1969 the Bulletin itself includes a keyword index and annual index-contents to the *Bulletin* are published in place of the *Guide*.

Another bibliography which ranges over all temperatures below 0 °C is *Cryogenics and refrigeration: a bibliographical guide*, by E. M. Codlin (Macdonald, 1968–   ). Two volumes have appeared so far— the first covering the years 1908–67 with the main emphasis after 1950, and the second covering the years 1966–68. It includes abstracts, a list of the periodicals cited and general sources of information. As well as references to the physics and engineering aspects of low temperatures, the *Guide* refers to applications in fields such as biology, medicine, electronics and space research.

Progress in low temperature physics (North-Holland) produced six volumes between 1955 and 1970, when it was decided to replace the series by individual monographs; *Progress in cryogenics* (Heywood) produced four volumes between 1959 and 1964. Of the current periodicals on the subject, *Journal of Low Temperature Physics* (Plenum) publishes papers on fundamental, theoretical and experimental research—it does not contain papers of a technical or applied nature. Included among its subjects are superfluidity and superconductivity, phase transitions, thermal properties, magnetism and surface phenomena at low temperatures. Another journal, *Cryogenics* (Heywood), has as its subtitle 'The International Journal of Low Temperature Engineering and Research'; it covers all aspects and includes research papers, technical notes, conference reports and news items.

Low temperature physics, by L. C. Jackson (5th edn, Methuen, 1962), provides a short monograph for undergraduates, with chapters on the production and measurement of low temperatures, liquid and solid helium, specific heats, electrical conductivity and magnetism. For a standard reference book there is *Cryogenic fundamentals*, edited by G. G. Haselden (Academic Press, 1971), which is a collection of papers and physical data covering the whole field of very low temperatures from liquid methane down to the lowest temperatures reached. *Advanced cryogenics*, edited by C. A. Bailey (Plenum, 1971), is based on a lecture course for students of a variety of sciences, and discusses the fundamental thermodynamics, properties of materials and applications. *Low temperature solid state physics: some selected topics*, by H. M. Rosenberg (Clarendon, 1963), considers specific heats, thermal and electrical conductivities, magnetic and mechanical properties, thermoelectricity, thermal expansion and the behaviour of super- and semiconductors.

A comprehensive book on experimental laboratory methods is *Experimental techniques in low-temperature physics*, by G. K. White (2nd edn, Clarendon, 1968). It discusses the production and measurement of low temperatures, the handling of liquefied gases on a laboratory scale, and the principles and design of experimental cryostats. A specific research technique which is used to reach temperatures below 1 K is described in *Principles and applications of magnetic cooling*, by R. P. Hudson (North-Holland, 1972). In fact nuclear magnetic cooling is capable of producing temperatures in the microkelvin region.

There are also several books which deal with the properties of substances at low temperatures. *Mechanical properties of materials at low temperatures*, by D. A. Wigley (Plenum, 1971), provides a thorough treatment of the properties of metals, polymers and composites. E. S. R. Gopal's *Specific heats at low temperatures* (Plenum, 1966) gives a fairly comprehensive survey of the subject, including both theory and experiment. The International Institute of Refrigeration sponsored a series of lectures at the University of Grenoble on 8 June 1965, which were later published as *Liquid Hydrogen* (*properties, production and applications*) (Pergamon, 1966). Finally, *Thermophysical properties of liquid air and its components*, by A. A. Vasserman and V. A. Rabinovich (Israel Program for Scientific Translations, 1970), analyses data on liquid air, oxygen, argon and nitrogen.

THERMOPHYSICAL DATA

In addition to the general collections of physical data, there are several which specialise in the thermophysical properties of substances. In particular, the Thermophysical Properties Research Center at Purdue University, Indiana has been responsible for the production of two very comprehensive publications—a literature guide and a compilation of data. *The thermophysical properties research literature guide*, edited by Y. S. Touloukian *et al.* (2nd edn, Macdonald, Plenum, 1967), covers references to the properties under seven headings: thermal conductivity, specific heat, viscosity, thermal radiative properties, diffusion coefficient, thermal diffusivity and Prandtl number. It was originally being published periodically and the first two volumes appeared in 1960 and 1963. Then it was decided to merge Volumes 1 and 2 with the proposed Volume 3 into a new edition which covers world literature (research papers, theses and reports) from 1920 (and some earlier material) to June 1964. The publication is in three separate books: the first contains necessary information for users of the *Guide* with other information such as a classified directory of substances for which information is given, and a dictionary of synonyms and trade names; the second contains the index covering the seven properties and a code which refers the user to the third book where bibliographic citations and an author index are to be found. It is intended to produce further updating volumes.

The other major publication from the Center is entitled *Thermophysical properties of matter—the TPRC Data Series* (Plenum, 1970–   ). To be completed in 13 volumes, the first 10 volumes have been published so far. Data is being presented under the same seven headings used for the *Literature guide*. The TPRC has a 'Data Update Plan' under which a subscriber can inquire for any new data which have been processed at the Center but not yet published, or for any revisions to published data. The Center also produces annual listings of corrigenda which will continue during the lifetime of an edition of the *Data Series*.

A book of tables intended for students which includes, among other things, abbreviated steam tables, enthalpies of various gaseous hydrocarbons and critical constants for several organic and inorganic substances, is *Abridged thermodynamic and thermochemical tables with charts (British units)*, by F. D. Hamblin (Pergamon, 1968). Abbreviated data can also be found in *Selected values of thermodynamic properties of metals and alloys*, by R. Hultgren *et al.* (Wiley, 1963). *Thermophysical properties of gases and liquids*, edited by V. A. Rabinovich (Israel Program for Scientific Translations, 1970), consists of a collection of translated papers originally published by the USSR Government Standards Office, and includes reports on theoretical and experimental work, and data in graphic and/or tabulated form.

Finally, two books which contain tables of radiation data. *Tables of blackbody radiation functions*, by M. Pivovonsky and M. R. Nagel (Macmillan, 1961), presents a tabulation of Planck's radiation law with a number of related functions and some auxiliary working tables. *Thermal radiative properties*, by W. D. Wood *et al.* (Plenum, 1964), includes data for a variety of metals and alloys, coated materials for high-temperature use, ceramics and graphite.

12

Light, electricity and magnetism

J. W. Burchell

The aim of this chapter is to survey the literature of light, electricity and magnetism, by indicating and describing the relevant abstracting journals and the most important reviews, journals and books which will provide the current state of knowledge and recent developments in these subjects. The use made of the literature will, naturally, depend very much on the needs of the individual—whether he is a student, an information worker or a research scientist checking on the latest advances in his field or seeking background knowledge for a new area of research. An attempt has been made to assist the reader by indicating the level and scope of the items mentioned.

Although we find it convenient to divide science up into a number of broad subjects, and this chapter is concerned with three of the 'classical' divisions of physics, we know only too well that the phenomena themselves do not always fit conveniently into the pigeon-holes we allot them. Anyone who has done library classification soon realises that the literature often does not match up easily with whatever system of classification is being used. As science has developed, subjects once thought to be quite separate have been discovered to be very much interconnected. Thus, after Faraday demonstrated the links between electricity and magnetism, Maxwell developed his electromagnetic wave theory. Many of the books of electromagnetic theory mentioned later, while mainly concerned with electric and magnetic fields, do include chapters on the propagation of light. As a result of this lack of clearly defined divisions, much of electricity, for example, could be considered as the province of the engineer, and some of magnetism might come under solid state physics. In this chapter the net has been spread as wide as possible, within the confines of the space allowed, to include as large a range of literature as possible.

LIGHT

The old saying that 'seeing is believing' may not be a good motto for the progress of science, but it does indicate the fundamental part that vision, and consequently light, plays in our understanding of the physical world, as well as in our daily lives. Although one of the oldest sciences, the study of optics has not stood still in recent years. Optical instruments such as the telescope, the camera, the microscope and the spectroscope have all played, and continue to play, an essential role in astronomy, biology and virtually every branch of science as well as physics itself. Optics and the study of light have been vital in the development of the modern theories of physics and cosmology such as relativity, quantum theory and the expanding universe. In recent years important advances in applied optics have included the laser, holography and optical fibres. A helpful layman's introduction to modern optics and most of these new advances is *Revolution in optics*, by S. Tolansky (Penguin, 1968).

General journals and abstracts

Journals dealing exclusively with specific aspects of light will be mentioned under those aspects, but there are several which cover the whole, or large sections of the subject. These are in addition to the general physics journals described elsewhere, which naturally include papers on optics. For review articles on the latest developments in all parts of optics there is the series *Progress in Optics* which began in 1961 and is published approximately annually by North-Holland. A cumulative index to the first 10 volumes is included in Volume 10 (1972). Review-type papers are also found in the proceedings of conferences and meetings of the International Commission for Optics which are published usually in various physics journals, although some do appear as monographs.

There is no abstracts journal devoted entirely to optics. The main bibliographical source in English is *Physics Abstracts*, in which the subject breakdown includes headings such as 'Optical Instruments and Techniques' (including lasers and photographic instruments), 'Optics', 'Quantum Optics' and, in the section on condensed matter, 'Optical Properties and Condensed Matter Spectroscopy'. *Bulletin Signalétique* (CNRS) includes optics in *Section 130* and spectroscopy in *Section 165*. *Technisches Zentralblatt* (Akademie) lists papers on all parts of physical and technical optics. Optics and spectroscopy are also included in *USSR and East Europe Scientific*

Abstracts, Physics and Mathematics, which is produced by the US Department of Commerce, Joint Publications Research Service. A certain amount on the applied aspects of the subject can be found in *Engineering Index* and (for British work) *British Technology Index*.

The following journals are major sources of research papers:

Applied Optics, published by the Optical Society of America. It covers all types of applications of facts, principles and methods of optics.

Atti della Fondazione Giorgio Ronchi and Contributi dell'Instituto Nazionale di Ottica (Florence) includes a considerable amount of physiological optics.

Jenaer Jahrbuch, published by VEB Gustav Fischer Verlag. Contains articles in German on theoretical and experimental optics.

Journal of the Optical Society of America covers the whole of optics from both the experimental and theoretical point of view, including spectroscopy, colorimetry, physiological optics and vision.

Nouvelle Revue d'Optique (Masson). In French with English titles, sections are included devoted to abstracts and recent patents. Mainly concerned with applied optics and instrumentation but it does contain some theoretical papers.

Optica Acta (Taylor and Francis). Its subtitle is 'The International Journal of Applied and Theoretical Optics', which is an indication of the width of its coverage.

Optics and Spectroscopy (AIP). A cover-to-cover translation of the Russian journal *Optika y Spektroskopiya*.

Optics and Communication (North-Holland). It is described as 'a journal devoted to the rapid publication of short contributions in the field of optics and interaction of light with matter'.

Optik—Zeitschrift für Licht und Elektronenoptik (Wissenschaftliche Verlagsgesellschaft, Stuttgart). Articles are in German or English with title and abstract in the other language. Covers practical optics, physiological optics, electron microscopy, measurement of optical properties, lasers, etc.

Science of Light. Published by the Institute for Optical Research, Kyôiku University, Tokyo, it contains a considerable amount of spectroscopy as well as some general optics.

General textbooks

There are several introductory textbooks for undergraduates which can also prove useful as reference works. One of the most popular of these is *Fundamentals of optics*, by F. A. Jenkins and H. E. White

(3rd edn, McGraw-Hill, 1957), and another excellent book is R. S. Longhurst's *Geometrical and physical optics* (2nd edn, Longmans, 1967). At a rather more advanced level there is *Principles of optics— electromagnetic theory of propagation, interference, and diffraction of light*, by M. Born and E. Wolf (4th edn, Pergamon, 1970). Other good texts include: *Modern optics*, by E. B. Brown (Reinhold, 1965); *Optics*, by F. G. Smith and J. H. Thomson (Wiley, 1971); and *University optics*, by D. W. Tenquist, R. M. Whittle, and J. Yarwood (Iliffe, 2 vols., 1969–70).

Geometrical optics and instrumentation

Rays of light may not really exist, but the techniques of ray-tracing still provide valuable information on the design of optical components and instruments. Most of the textbooks mentioned above include chapters on geometrical optics. Some more specialised monographs follow:

Fry, G. A. *Geometrical optics*. Chilton, 1969. Written from the traditional ray-tracing point of view.

Stravroudis, O. N. *The optics of rays, wavefronts and caustics*. Academic Press, 1972. A detailed study of the mathematical foundations of geometrical optics, including some novel concepts and techniques.

Zimmer, H. G. *Geometrical optics*. Springer, 1970. An introduction requiring a knowledge of elementary geometry and algebra; the subject is developed from the principle of the conservation of radiated energy.

Two books which demonstrate that geometrical optics can be considered as an approximation to Maxwell's electromagnetic theory are: *Electromagnetic theory and geometrical optics*, by M. Kline and I. W. Kay (Interscience, 1964), and *Mathematical theory of optics*, by R. K. Luneberg (University of California Press, 1964). A specialised book which contains a Hamiltonian treatment of aberration theory is *An introduction to Hamiltonian optics*, by H. A. Buchdahl (CUP, 1970). A mathematical treatment of the subject is found in *Modern geometrical optics*, by M. Herzberger (Interscience, 1958). This book, which is intended to be of practical use to designers of optical instruments, develops a mathematical model of an optical system from which all characteristics of the optical image can be obtained.

As mentioned already in the introduction to this section, optical instruments play an essential part in all branches of science. Because of the inherent limitations of such instruments and of human vision, it is essential for the instruments to be designed as carefully as possible to obtain maximum benefit from them. An introductory book for prospective instrument designers is *Instrumental optics*, by G. A. Boutry (Hilger and Watts, 1961). At a more advanced level there is *Modern optical engineering: the design of optical systems*, by W. J. Smith (McGraw-Hill, 1966), and a comprehensive book which includes the application of computers is *A system of optical design*, by A. Cox (Focal Press, 1964). The latter includes a bibliography and an appendix in which nearly 300 lens designs have been recomputed using a specially devised program, and evaluated with graphs to show their performance. One further book which contains a considerable amount of numerical tables, some specially computed, is *Applied optics—a guide to optical systems design*, by L. Levi (Wiley, Vol. 1, 1968). Three books concerned particularly with lens design, and published by Adam Hilger in their series *Monographs on Applied Optics* are: *The optical transfer function*, by K. R. Barnes (1971); *Optimization techniques in lens design*, by T. H. Jamieson (1971); and *Lens aberration data*, by J. M. Palmer (1971).

For descriptions of optical instruments and their practical use the following books are recommended:

Kingslake, R. (ed.). *Applied optics and optical engineering*. 5 vols. Academic Press, 1965–69. Includes volumes on the detection of light and infra-red radiation, optical components, and (two volumes) on optical instruments.

Martin, L. C. *Technical optics*. 2nd edition. Pitman, Vol. 1, 1966; Vol. 2, 1961. An introductory text—the first volume deals with some general optics and includes a short introduction to the theory of spectacles; the second volume describes a number of instruments.

Van Heel, A. C. S. *Advanced optical techniques*. North-Holland, 1967

The published proceedings of two conferences are also valuable sources of information on new instruments and techniques:

Dickson, J. H. (ed.). *Optical instruments and techniques*. Oriel Press, 1970. Conference held at the University of Reading, July, 1969.

Habell, K. J. (ed.). *Proceedings of the Conference on Optical Instruments and Techniques, London 1961*. Chapman and Hall, 1962

While the physicist is unlikely to be concerned with the actual manufacture of optical components, it will be useful to note two books which are concerned with this aspect. A general text is D. F. Horne's *Optical production technology* (Adam Hilger, 1972) and a more specific, mainly theoretical book is *Generation of optical surfaces*, by K. G. Kumanin (The Focal Library, 1967).

There are several journals published by the manufacturers of optical instruments, mostly German in origin, such as *Jena Review, Leitz Mitteilungen* and *Zeiss Information*. For some of these there are separate German and English editions. Other journals include *Optical Engineering* (the journal of the Society of Photo-Optical Instrumentation Engineers), *Soviet Journal of Optical Technology* (ATP, a cover-to-cover translation of *Optiko-Mekhanicheskaya Promyshlennost*). The Society of Photo-Optical Instrumentation Engineers also publishes their *Seminar Proceedings*, each of which deals with a specific topic, and the *Proceedings of the Annual Technical Symposium*.

Diffraction, interference, scattering

Of importance in the development of the wave theory of light, these phenomena have significant practical applications in fields such as spectroscopy, crystallography and the production of optical components.

An introductory text based on an undergraduate course on Fraunhofer diffraction is *Diffraction: coherence in optics*, by M. Françon (Pergamon, 1966), and a fairly advanced work on Fraunhofer diffraction patterns is *Optical transforms*, edited by H. Lipson (Academic Press, 1972). This includes chapters on their applications, such as the determination of crystal structure, biological studies, optical data processing and holography. A mathematical treatment of those parts of diffraction theory which have immediate practical applications is found in *Applied diffraction theory*, by F. H. Northover (American Elsevier, 1971); this book requires a knowledge of classical analysis and electromagnetic theory, although the mathematical techniques and methods used are presented in full: it includes examples involving micro- and radiowaves. There are two books by J. Guild on the theory and use of diffraction gratings—*The interference systems of crossed diffraction gratings: theory of Moiré fringes* (OUP, 1956), and *Diffraction gratings as measuring scales: practical guide to the metrological use of Moiré fringes* (OUP, 1960).

An important starting point for the study of interferometry is *An introduction to interferometry*, by S. Tolansky (2nd edn, Longmans, 1973). A further two books which, in this case, are concerned with the practical use of interferometry are: *Interferometry as a measuring tool* by J. Dyson (Machinery Publishing Co., 1970), and *Polarization interferometers: applications in microscopy and macroscopy*, by M. Françon and S. Mallick (Wiley/Interscience, 1971). In both of these the mathematics is kept to a minimum and the emphasis is on physical principles and practical uses. The technique of multiple beam interferometry is used to study the microstructure of surfaces; the theory and practice is explained in *Surface microtopography*, by S. Tolansky (Longmans, 1960).

The effect of the scattering of light is familiar to everyone because it is responsible for the blue colour of the sky. H. C. Van de Hulst has written *Light scattering by small particles* (Chapman and Hall, 1957), which contains the theory for various types of particle. *Molecular scattering of light*, by I. L. Fabelinskii (Plenum, 1968) describes more recent theoretical and experimental research on molecular scattering in gases, liquids and solids (including glasses and crystals), including methods for studying scattered light and its spectral composition, and the use of different light sources. Another book which develops the theory of scattering by spheres and cylinders, gives approximations for other shapes, discusses applications and includes the scattering of micro- and radiowaves is *The scattering of light and other electromagnetic radiation*, by M. Kerker (Academic Press, 1969). Recent advances in Raman and Brillouin scattering in particular are found in the *Proceedings of the International Conference on Light Scattering in Solids*. The first was held in New York in 1968 and published by Springer in 1969; the second in Paris in 1971 and published by Flammarion Sciences in the same year.

The velocity of light

This is a quantity which holds a fundamental place in modern physics, and a considerable amount of ingenuity has gone into its accurate measurement. A book which examines and compares the various methods used is *The velocity of light and radio waves*, by K. D. Froome and L. Essen (Academic Press, 1969). Modern methods, including the use of high frequency radio waves, are fully described.

Quantum optics

The discovery that radiation interacts with matter in a manner that can only be explained by considering the radiation to be composed of discrete particles, labelled photons, is one of the foundations of quantum theory. A fairly recent student text in which the emphasis is on quantum optics is *States, waves and photons: a modern introduction to light* by J. W. Simmons and M. J. Guttmann (Addison-

Wesley, 1970); a basic knowledge of matrix algebra and electromagnetic theory is required for this book. A more advanced book is *Fundamentals of quantum optics*, by J. R. Klauder and E. C. G. Sudarshan (Benjamin, 1968). It includes an account of the quantum theory of partial coherence and the relationship between classical and quantum theories of coherence. Another advanced book which contains a comprehensive account of the subject is *The quantum theory of light*, by R. Loudon (OUP, 1973). Finally, *Optical interactions in solids*, by B. di Bartolo (Wiley, 1968), provides a theoretical background for research workers concerned with lasers and absorption and fluorescence spectroscopy of solids.

Lasers, coherence, holography and non-linear optics

Of all modern optical instruments, the laser has probably received the most attention, both in popular imagination and among scientists. The basic principle was elaborated in 1958, the first working laser was produced in 1960, and since then there has been a rapidly increasing number of papers. So many appear, in fact, that the subject has merited a monthly abstracts journal of its own entitled *Journal of Current Laser Abstracts* (Monthly. Institute for Laser Documentation, Felton, California). The following are some of the bibliographies:

Kamal, A. K. *Laser abstracts*. Plenum, 1964. Covers the period from the earliest research to mid-1963.

Tomiyasu, K. *The laser literature: an annotated guide*. Plenum, 1968. Begins where the previous volume leaves off and covers literature published up to the end of 1966.

Ashburn, E. V. *Laser literature: a permuted bibliography, 1958–1966*. 2 vols. Western Periodicals Co., 1967. The year 1967 is covered by a supplement published in 1968.

There are many good introductory texts on lasers; the following is a selection from those published in the last few years:

Brown, R. *Lasers: a survey of their performance and applications*. Business Books, 1969. A popular, non-mathematical approach.

Lengyell, B. A. *Lasers*. 2nd edition. Wiley/Interscience, 1971. Another book at undergraduate level.

Maitland, A. and Dunn, M. H. *Laser physics*. North-Holland, 1969. An introduction which adopts a semiclassical approach.

Melia, T. P. *An introduction to masers and lasers*. Chapman and Hall, 1967

For those who require more specialised information on lasers, the following books are recommended:

Arecchi, F. T. and Schulz-Dubois, E. O. (eds.). *Laser handbook* 2 vols. North-Holland, 1972. An encyclopaedic review of all aspects of lasers and their applications which is intended for those who already have a background knowledge.

Maitland, A. and Dunn, M. H. *Laser physics*. North-Holland, 1969. A theoretical treatment for graduates in physics or electronic engineering.

Marshall, S. L. (ed.). *Laser technology and applications*. McGraw-Hill, 1968. In addition to the theory, the various types of lasers and laser materials are described in some detail.

Pressley, R. J. *Handbook of lasers with selected data on optical technology*. Chemical Rubber Co., 1971. A collection of critically evaluated original data relevant to laser research and development.

Ratner, A. M. *Spectral, spatial and temporal properties of lasers*. Plenum Press, 1972. A translation from the Russian; it is concerned only with solid state lasers with optical pumping.

Röss, D. *Lasers, light amplifiers and oscillators*. Academic Press, 1969. The German edition of 1966 has been updated in this English translation. It contains a bibliography of over 4000 references and also tables of the main characteristics of more than 1000 laser transitions in gases, and the important material parameters of all known crystal and semiconductor lasers.

There are also several monographs which deal with specific types of laser; again only a selection of these can be given:

Allen, L. and Jones, D. G. C. *Principles of gas lasers*. Butterworths, 1967. For postgraduate physicists.

Gooch, C. H. *Gallium arsenide lasers*. Wiley/Interscience, 1969. Covers most aspects from the preparation and properties of gallium arsenide to the applications of the devices.

Patek, K. *Glass lasers*. Iliffe, 1970. Discusses solid state lasers with a glass or crystal active element and optical pumping. Contains a large number of tables on the properties of glasses.

Rieck, H. *Semiconductor lasers: basic physics, technology and design*. Macdonald, 1970

The use of lasers in plasma physics is the subject of *Laser interaction and related plasma phenomena. Proceedings of the first workshop held at Rennsselaer Polytechnic Institute, Connecticut, 9–13 June, 1969*, edited by H. J. Schwarz and H. Hora (Plenum Press, 1971). A laboratory reference book which includes the techniques for measuring the energy, power gain, wavelength, etc., of laser beams is *Laser parameter measurements handbook*, edited by H. G. Heard (Wiley, 1968).

Optical coherence has been studied for a long time, but the development of the laser with its highly concentrated beam of coherent light has increased interest in the subject in recent years. Two introductory books are *An introduction to coherent optics and holography*, by G. W. Stroke (2nd edn, Academic Press, 1969), and *Optical coherence theory—recent developments*, by G. J. Troup (Methuen, 1967). A comprehensive survey written from the electrical engineering viewpoint and concerned mainly with lasers and associated devices such as modulators, antennas and receivers is A. F. Harvey's *Coherent light* (Wiley/Interscience, 1970). For those requiring a theoretical and mathematical treatment there is *Coherence of light*, by J. Perina (Van Nostrand Reinhold, 1972).

The theory of holography was developed in the late 1940s but it was not until the coherent light of the laser was available that it became a practical possibility. As with the laser itself, a great deal has been written on the theory and use of holography in the last 10 years. Most of the abstracts journals already mentioned include papers on holography. In addition, *Photographic Abstracts*, published by the Royal Photographic Society of Great Britain, contains a section on holography in each of its monthly issues.

The following books have been selected from the many that have been published:

Butters, J. N. *Holography and its technology*. Peregrinus, 1971

Camatini, E. (ed.). *Optical and acoustical holography. Proceedings of the NATO Advanced Study Institute, Milan, Italy, 24 May–4 June, 1971*. Plenum Press, 1972. Covers theory and applications of optical, microwave, and acoustical holography.

Collier, R. J. and others. *Optical holography*. Academic Press, 1971. Covers the theory and practice of producing holograms and explains various applications, including information storage. Includes chapters on colour holography and computer-generated holograms.

Lehmann, M. *Holography—technique and practice*. Focal Press, 1970. A practical book intended for the laboratory engineer or technician.

Under certain circumstances, such as an increase in the intensity of an incident light beam, certain materials show optical non-linearity. This is another area of research which has been affected by the development of the laser. An elementary but thorough grounding in the subject is found in *An introduction to nonlinear optics*, by G. C. Baldwin (Plenum, 1969). Two more advanced books are *Problems of nonlinear optics*, by S. A. Akhmanov and R. V. Khoklov (Gordon and Breach, 1972), and *Applied Nonlinear optics*, by F. Zernike and J. E. Midwinter (Wiley, 1973). The first of these is a translation from the Russian and summarises a lot of work on the theory of non-linear media up to 1963. The second is intended for physicists and engineers interested in the applications of non-linear devices.

Fibre optics and optical communication

Another remarkable development in optics in recent years is the ability to 'pipe' an optical image along a bundle of fibres which can be twisted as the situation demands. A comprehensive book on the subject which covers basic principles, theory, technology and its applications is *Fibre optics: principles and applications*, by N. S. Kapany (Academic Press, 1967). A useful bibliography which contains over 900 entries covering the period January 1960 to June 1971 is *A bibliography of fibre optics* (Hampshire Technical Research Industrial Commercial Service, Southampton, 1971).

The principles and methods of light transmission which are po-tentially useful for optical communications systems are described in *Light transmission optics*, by D. Marcuse (Van Nostrand Reinhold, 1972). It includes chapters on optical fibres and various types of optical waveguides. *Integrated optics*, by D. Marcuse (IEEE Press, 1973), includes 35 reprinted papers dealing with the theory and use of thin-film waveguides. The application of laser light to optical communications systems is discussed in *Laser receivers: devices, techniques, systems*, by M. Ross (Wiley, 1966). *Optical waveguides*, by N. S. Kapany and J. J. Burke (Academic Press, 1972), describes waveguide phenomena in classical optical terms as far as possible, having been written for the optical scientist rather than the micro-wave engineer.

Spectroscopy

Our present knowledge of atomic physics owes much to spectroscopic studies. A considerable amount of the literature, however, has been written for the chemist, for whom spectroscopy is an important analytical tool. We must concentrate here on the physics literature, but some books will be mentioned which are mainly intended for chemists. Also, the periodicals which follow are mostly analytical chemistry in content, although they do contain articles on physical principles as well.

Review articles are to be found in the annual series *Applied*

Spectroscopy Reviews (Arnold), which covers the entire range from X-ray to radiofrequency spectroscopy. The bi-monthly journal *Applied Spectroscopy* (Society for Applied Spectroscopy, Baltimore) publishes articles on the theory and practice of absorption, emission, Raman, mass and nuclear magnetic resonance spectroscopy. For Japanese research there is *Bunkô Kenkyû: Journal of the Spectroscopical Society of Japan*; it is in Japanese but contains titles, and often abstracts, in English. Russian work is reported in *Journal of Applied Spectroscopy* (Plenum), which is a cover-to-cover translation of *Zhurnal Prikladnoi Spektroskopii*. *The Journal of Quantitative Spectroscopy and Radiative Transfer* (Pergamon) is concerned with topics such as theoretical and experimental studies of spectral line intensities and shapes, emissivities of heated materials, long-range detection of radiating sources, etc.

The entire field of spectroscopy is surveyed in *The encyclopedia of spectroscopy*, edited by G. J. Clark (Reinhold, 1960). It contains articles on the different types of spectroscopy, instrumentation and applications. There have been many developments since this encyclopaedia was published, but it remains a valuable introduction.

There follows a selection of books which are concerned with the relationship between observed spectra and the structure of atoms and molecules. The classical methods for analysing the properties of atoms are found in *The theory of atomic spectra*, by E. U. Condon and G. H. Shortley (CUP, 1935), and a general account of the subject including many later developments has been given by J. C. Slater in *Quantum theory of atomic structure* (McGraw-Hill, 1960). C. Candler's *Atomic theory and the vector model* (2nd edn, Hilger and Watts, 1964) relates observations to the vector model of the atom, and leaves out the difficult mathematics as far as possible. The Bohr–Sommerfeld vector model is also developed in *Principles of atomic spectra*, by B. W. Shore (Wiley, 1968), which then goes on to consider the mathematical foundations of wave mechanics and quantum theory and to extend the theory of atomic structure and spectra. The physical structure of the atom is discussed in *Atomic spectra*, by H. G. Kuhn (2nd edn, Longmans, 1969), and an advanced treatment of the physical principles and theory of atomic spectroscopy is found in *Introduction to the theory of atomic spectra*, by I. I. Sobel'man (Pergamon, 1972). *Spectroscopy and structure*, by R. N. Dixon (Methuen, 1965), is intended for honours degree students and presents the physical bases of a number of mathematical models of atoms and molecules. An introductory book on the various types of atomic and molecular spectroscopy and the theoretical determination of energies and structures is *Spectroscopy and molecular structure*, by G. W. King (Holt, Rinehart and Winston, 1964).

Many of the following books and serials, which describe different types of spectroscopy, have been written for the analytical chemist, but do contain some discussion of the basic physical principles involved:

Absorption spectroscopy

Dean, J. A. and Rains, T. C. *Flame emission and atomic absorption spectrometry*. Dekker, to be completed in 3 vols. Vol. 1, 1969; Vol. 2, 1971.

Lothian, G. F. *Absorption spectrophotometry*. 3rd edition. Adam Hilger, 1969

Slavin, W. *Atomic absorption spectroscopy*. Interscience, 1968. A critical guide to the literature.

Fluorescent spectroscopy

Guilbault, G. G. *Practical fluorescence: theory, methods and techniques*. Dekker, 1973.

Pesce, A. J. *et al.* (eds.). *Fluorescence spectroscopy: an introduction for biology and medicine*. Dekker, 1971. Includes theory, and in spite of title should prove a useful guide to experimental procedures for physicists and chemists.

Transform spectroscopy

Bell, R. J. *Introductory Fourier transform spectroscopy*. Academic Press, 1972. Includes an extensive bibliography.

Mertz, L. *Transformations in optics*. Wiley, 1965. Mainly concerned with Fourier transform spectrometry, but does consider Fresnel transformations as well.

Raman spectroscopy

Advances in Raman Spectroscopy, Vol. 1, published in 1973, consists of the proceedings of the 3rd International Conference on Raman Spectroscopy, University of Reims, September, 1972.

Anderson, A. *The Raman effect*. 2 vols. Dekker, 1971

Koningstein, J. A. *Introduction to the theory of the Raman effect*. Reidel, 1972.

Szymanski, H. A. (ed.). *Raman spectroscopy: theory and practice.* Plenum, Vol. 1, 1967; Vol. 2, 1970

Tobin, M. C. *Laser Raman spectroscopy.* Wiley/Interscience, 1971. Particular emphasis is laid on experimental techniques and recent developments in instrumentation.

Reflectance spectroscopy

Harrick, N. J. *Internal reflection spectroscopy.* Interscience, 1967. Covers theory, instrumentation and applications.

Wendlandt, W. W. (ed.). *Modern aspects of reflectance spectroscopy. Proceedings of the American Chemical Society Symposium on Reflectance Spectroscopy, 11–12 September 1967.* Plenum, 1968

Gamma-ray spectroscopy

Crouthamel, C. E. *Applied gamma-ray spectrometry.* 2nd edition. Pergamon, 1970

Quittner, P. *Gamma-ray spectroscopy with particular reference to detector and computer evaluation techniques.* Adam Hilger, 1972. Describes the use of computers in modern scintillation and semi-conductor gamma-ray spectrometry.

Shafroth, S. M. *Scintillation spectroscopy of gamma radiation.* Gordon and Breach, 1967

X-ray spectroscopy

Jenkins, R. and De Vries, J. L. *Practical X-ray spectrometry.* 2nd edition. Macmillan, 1970.

Ultra-violet spectroscopy

Jaffé, H. H. and Orchin, M. *Theory and applications of ultraviolet spectroscopy.* Wiley, 1962

Infra-red spectroscopy

Chantry, G. W. *Submillimeter spectroscopy: a guide to the theoretical and experimental physics of the far infra-red.* Academic Press, 1971

Miller, R. G. J. and Stace, B. C. (eds.). *Laboratory methods in infra-red spectroscopy.* 2nd edition. Heyden, 1972

Möller, K. D. and Rothschild, W. G. *Far-infra-red spectroscopy.* Wiley/Interscience, 1971

Stewart, J. E. *Infra-red spectroscopy: experimental methods and techniques.* Dekker, 1970

Microwave spectroscopy

Ingram, D. J. E. *Spectroscopy at radio and microwave frequencies.* 2nd edition. Butterworths, 1967

Rebane, K. K. *Impurity spectra of solids: elementary theory of vibrational structure.* Plenum, 1970

Sherwood, P. M. A. *Vibrational spectroscopy of solids.* CUP, 1972

Wollrab, J. E. *Rotational spectra and molecular structure.* Academic, 1967. A review of some of the areas covered by microwave spectroscopy; contains an extensive bibliography.

There are many books and large collections (e.g. the *Sadtler Spectra*) of spectral data. The Science Reference Library in London holds a wide range of these and has published a helpful guide to its stock, *A guide to the literature on special data,* by A. Clarke (2nd edn, 1972), which serves as an annotated bibliography for those seeking such data.

Finally, a few books on spectroscopic techniques and instrumentation:

Blackburn, J. A. (ed.). *Spectral analysis: methods and techniques.* Dekker, 1970. For these who have to interpret spectral data of any kind, optical or otherwise; covers mathematical techniques and several applications.

Bousquet, P. *Spectroscopy and its instrumentation.* Adam Hilger, 1971

Davis, S. P. *Diffraction grating spectrographs.* Holt, Rinehart and Winston, 1970

Ultra-violet and infra-red radiation

It is convenient to introduce at this point some of the literature on these parts of the electromagnetic spectrum which are not visible to the human eye, as mention has just been made of books of ultra-violet and infra-red spectroscopy.

The ultra-violet region had received less attention until recent years, when space research caused a greater interest in ultra-violet spectroscopy. A book directly connected with this new interest is

The middle ultraviolet: its science and technology, edited by A. E. S. Green (Wiley, 1966). It deals with radiation of wavelength 3400 to 1700Å, which is the region excluded from ground-based observers by atmospheric ozone and oxygen. The various sources of ultra-violet, including solar radiation, and the properties of materials that reflect and transmit it are described in *Ultraviolet radiation*, by L. R. Koller (2nd edn, Wiley, 1965). An introduction to the applications of both infra-red and ultra-violet is provided by *Ultra-violet and infra-red engineering*, by W. Summer (Pitman, 1962).

The quarterly periodical *Infra-red Physics* (Pergamon) covers all aspects of the subject, especially spectroscopy and infra-red detectors. *Infra-red physics*, by J. T. Houghton and S. D. Smith (OUP, 1966), also emphasises spectroscopy. Particularly useful for its collection of data and formulae is M. A. Bramson's *Infra-red radiation: a handbook for applications with a collection of reference tables* (Plenum, 1968). The infra-red is used in a variety of ways, and many of these are described in *Practical application of infra-red techniques*, by R. Vanzetti (Wiley, 1972).

Polarisation of Light

An outline of the theory, a description of the various types of polariser and a wide range of applications are to be found in *Polarized light: production and use*, by W. A. Schurcliff (Harvard University Press, 1962). A second book written for research workers and postgraduate students is *Polarized light and optical measurement*, by D. Clarke and J. F. Grainger (Pergamon, 1971), which also discusses the theory of polarisation and goes on to consider its role in optical measurement. *The physical basis of polarized emission*, by P. P. Feofilov (Consultants Bureau, 1961), is mainly concerned with the use of polarisation in studying the structure of a radiating system. Using the term 'depolarisation' to denote the change of an electromagnetic wave from one state of polarisation to another, an investigation of the theory of this change is found in *The depolarization of electromagnetic waves*, by P. Beckmann (The Golden Press, 1968).

Luminescence

This is a term which covers several phenomena involving the emission of light that cannot be attributed only to the temperature of the radiating body. The *Journal of Luminescence* (North-Holland) is devoted to all luminescent phenomena and materials; it contains many papers on materials in particular. Apart from that journal we shall ignore phenomena such as bioluminescence and chemiluminescence, and mention some of the literature on fluorescence, phosphorescence and electroluminescence.

In spite of its title, R. A. Passwater's *Guide to fluorescence literature* (Plenum) includes phosphorescence as well as fluorescence. Volume 1 was published in 1967 and contains nearly 5000 references covering the period 1950 to 1964. Volume 2, published in 1970, updates the first volume and covers the period late-1964 to mid-1968. The guide contains bibliographical details and a reference to the abstract in *Chemical Abstracts*, as well as author and subject indexes.

A survey of the whole field of photoluminescence excluding practical applications is to be found in *Fluorescence and phosphorescence*, by P. Pringsheim (Interscience, 1949). A more up-to-date and specialised account of the classical and quantum theories is given by B. I. Stepanov and V. P. Gribovskii in *Theory of luminescence* (Iliffe, 1968).

An introductory account of electroluminescence (the direct conversion of electrical energy into light) is to be found in *Electroluminescence*, by H. K. Henisch (Pergamon, 1962), which also includes a bibliography from 1891 to 1961. Later developments and a comprehensive treatment of the injection laser are described in P. R. Thornton's *The physics of electroluminescent devices* (Spon, 1967). Electroluminescent diodes and lasers are the subject of the specialist monograph *Injection electroluminescent devices*, by C. H. Gooch (Wiley, 1973).

Colour

An excellent journal which includes physical and physiological aspects of colour is *Die Farbe* (Musterschmidt-Verlag, Göttingen); articles are in German, English or French with abstracts in all three languages. The *Journal of Color and Appearance* contains items of news as well as articles on materials, applications, spectrophotometry, colorimetry, physiology, psychology, etc.

A useful introductory book which presents the fundamental information on the physics of colour, colour vision, colorimetry, and so on is *Color: a guide to basic facts and concepts*, by R. W. Burnham and others (Wiley, 1963). A comprehensive collection of physical data for both the theoretician and the practical worker to be found in *Color science: concepts and methods, quantitative data and formulas*, by G. Wyszecki and W. S. Stiles (Wiley, 1967). It includes details of light sources, filters and detectors; accounts of colour-matching, colour discrimination and adaptation; tables of the properties of human colour vision, etc.

There are several systems of colour identification—the Munsell system is a well-known one. The *Munsell book of colour* (Munsell Color Co., Baltimore, 1969) is offered in two collections with colour chips in either a glossy finish or a matt finish. A pocketbook which covers the basic facts of colour practice and includes a dictionary of colour samples with the British Standard and Munsell equivalents is *Methuen handbook of colour*, by A. Kornerup and J. H. Wonscher (2nd edn, Methuen and Co., 1967).

An outline of the properties of colour and its measurement is given in a book intended particularly for those concerned with colour photography: *Colour measurement*, by H. Arens (Focal Press, 1967). On the same subject, *The measurement of colour*, by W. D. Wright (4th edn, Adam Hilger, 1969), describes the trichromatic system of colour measurement.

Photometry and illumination

Much that is written on light measurement, sources and design is intended for architects and engineers, but we will close this section on the literature of light with some journals and books that include the physical aspects of this part of the subject. Two periodicals which cover the whole field, theoretical and applied, are *Lighting Research and Technology*, published by the Illuminating Engineering Society (London) and including an abstracts section; and *Journal of the Illuminating Engineering Society* (New York). The latter Society has also published the *IES lighting handbook*, edited by J. E. Kaufman (4th edn, 1966), which contains sections on the basic physics of light, vision, light measurement, sources and lighting appropriate for a wide variety of situations.

Devoted to the measurement of light, *Photometry*, by J. W. T. Walsh (3rd edn, Constable, 1958), is rather old now but still includes much useful information on practical methods. *Light calculation and measurements*, by H. A. E. Keitz (2nd edn, Macmillan, 1971), has the explanatory subtitle 'an introduction to the system of quantities and units in light-technology and to photometry'. A specialised method of photometry which uses lines or curves of equal density produced on a photographic plate is explained in *Equidensitometry: methods of two-dimensional photometry principles and fields of application*, by E. Lau and W. Krug (Focal, 1968), which includes a bibliography on the technique.

ELECTRICITY

To cover the whole literature of electricity (with, in addition, magnetism) in half a chapter is obviously an impossible task when one considers the possible extent of the subject. It deserves several chapters, if not a whole book, to itself. Admittedly a very large proportion of the literature is written by and for electrical and electronics engineers, and it might be argued that this has no place in a book on the literature of physics. But it is in practice impossible to make a clear division between physics and engineering literature, and a considerable amount of basic physical theory appears in engineering publications. Furthermore there is sufficient truth to matter in the view of one professor of electrical engineering who queried the point of studying physics when so many physicists later turn into electrical engineers! Be that as it may, the following pages must necessarily contain only a selection of all the literature connected in any way with the broad subject 'electricity', with, it is hoped, emphasis on the basic physics.

Abstracts journals

The major abstracting services in English covering the whole of electricity and magnetism are *Series A* and *Series B* of *Science Abstracts*. The first of these—*Physics Abstracts*—includes references to the electrical and magnetic properties of condensed matter. The second—*Electrical and Electronics Abstracts*—covers the whole field both pure and applied, including magnetic materials and devices, and electromagnetism. *Bulletin signalétique* (CNRS) has relevant abstracts in *Part 140: Electrotechnique*, which includes general electricity and magnetism, materials (conductors, dielectrics, etc.), electrical machines and power; and in *Part 145: Electronique*, which includes electronic measurement and devices, circuits, telecommunications and plasma physics. The German bibliographical journal *Technisches Zentralblatt, Abteilung Elektrotechnik* merged with other *Abteilungen* in 1971 to form the single series *Technisches Zentralblatt*. The earlier series dealt with the whole of electricity and magnetism, and included abstracts. The new title includes all of science and technology, but does not contain abstracts; it gives bibliographical details and descriptive keywords only. The main sections of *Referativnyi Zhurnal* (Viniti) which should be mentioned here are *Fizika* and *Elektrotekhnika i Energetika*. Abstracts of Soviet and East European work can also be found, in English, in *USSR and East Europe Scientific Abstracts, Electronics and Electrical Engineering* (Joint

Publications Research Service of the US Department of Commerce). *Engineering Index* can also be a useful guide to the literature, and does include references and abstracts for theoretical as well as practical articles.

General journals

The quantity of journals on electricity in all its aspects is very large indeed. Because of this, articles of interest to physicists are scattered widely and the use of abstracts journals becomes even more essential when making a subject search. The journals in the following list have been chosen because they represent those that contain a significant amount of basic physical theory, in addition to papers on applications and engineering.

Archiv für Elektrotechnik (Springer). Contents include both electrical and electronics engineering. Papers are mostly in German; some are in English; and all have an English title and abstract.

Proceedings of the Institute of Electrical and Electronics Engineers. Several issues are devoted to particular topics, but each one also contains a section of short communications on electronics, electro-magnetics, circuit theory, optics and quantum electronics, etc. (There are also several series of *IEEB Transactions* and some of these will be mentioned under more specific subject headings.)

Proceedings of the Institution of Electrical Engineers (London). Contents are divided into four sections—electronics; power control and automation; science; education and management.

Revue Générale de l'Electricite (Paris). Covers electricity, electro-technology and electronics.

Scientia Electrica (Basel). Concerned with 'modern problems of theoretical and applied electrical engineering'. Each issue contains two or three articles in German or English.

Soviet Electrical Engineering (Faraday). Cover-to-cover translation of *Elektrotekhnika*.

Handbooks and dictionaries

One standard reference work which contains a wide range of useful information on electrical units, materials, measurements and energy conversion is *Electrical engineer's reference book*, edited by M. G. Say (13th edn, Newnes-Butterworth, 1973). There are several dictionaries of electrical terms. Each definition in the *IEEE Standard Dictionary of Electrical and Electronic Terms* (Wiley/Interscience,

1972) is an official IEEE standard, and while American usage is stressed, the recommendations of the International Electrotechnical Commission have been adopted in many cases. Another useful and comprehensive dictionary is *DEIA. Dictionary of electrical abbreviations, signs and symbols*, by D. D. Polon (The Odyssey Press, 1967). A helpful multi-language dictionary which gives electrical terms in English/American, French, Spanish, Italian, Dutch and German is *Elsevier's electrotechnical dictionary in six languages*, compiled by W. E. Clason (Elsevier, 1965).

General textbooks

A popular textbook intended for the general degree-level student is *Electricity and magnetism*, by C. J. Smith (3rd edn, Edward Arnold, 1963). Another comprehensive and valuable text at a somewhat higher level (honours degree and first-year graduate) is *Electricity and magnetism*, by B. I. Bleaney and B. Bleaney (2nd edn, Clarendon, 1965). A third text of a similar nature is W. J. Duffin's *Electricity and magnetism* (2nd edn, McGraw-Hill, 1973), and a classic work which has been reprinted many times and retains its value is James Jeans's *The mathematical theory of electricity and magnetism* (5th edn, CUP, 1925, reprinted 1966). *Electricity and magnetism*, by J. Yarwood (University Tutorial Press, 1973), is a good student text.

Electromagnetism and electrodynamics

From those books which discuss electricity and magnetism in general we pass to the literature concerned with the interaction between them (electromagnetism, electromagnetic fields and waves) and between electrical, magnetic and mechanical phenomena (electrodynamics). Anyone seeking a textbook on electromagnetic theory is faced with a considerable number to choose from, written at different levels and from various points of view. Any selection of books is bound to be somewhat arbitrary, and those which follow are only intended to be typical of the more recently published ones.

Two introductory texts are *An introduction to electromagnetic theory*, by P. C. Clemmow (CUP, 1973), and *Fundamentals of electricity and magnetism*, by A. F. Kip (McGraw-Hill, 1969). A more advanced book which is largely mathematical and requires a previous conventional course in electromagnetism is F. N. H. Robinson's *Macroscopic electromagnetism* (Pergamon, 1973); the basic concepts and laws are discussed together with their relation to the atomic structure of matter. There are several books which develop electromagnetic theory from the special theory of relativity, among them *Electromagnetics*, by R. S. Elliott (McGraw-Hill, 1966), and *Classical electromagnetism via relativity*, by W. G. V. Rosser (Butterworths, 1968).

Many of the more advanced texts place an emphasis on field theory. The following have been selected as typical of these:

Becker, R. *Electromagnetic fields and interactions* 2 vols. Blackie, 1964. The first volume has the subheading 'Electromagnetic theory and relativity'; and the second, 'Quantum theory of atoms and radiation'.

Clemmow, P. C. *The plane wave spectrum representation of electromagnetic fields*. Pergamon, 1966. Explains how general electromagnetic fields can be represented by the superposition of plane waves travelling in various directions.

Eyges, L. *The classical electromagnetic field*. Addison-Wesley, 1972. A postgraduate text.

Good, R. H. and Nelson, T. J. *Classical theory of electric and magnetic fields*. Academic Press, 1971. An advanced text which includes electrostatics, magnetostatics, wave propagation and radiation.

Landau, L. D. and Lifschitz, E. M. *The classical theory of fields* 3rd English edition. Pergamon, 1971. Mainly concerned with electromagnetic fields.

Some of these books contain chapters on the theory of electromagnetic radiation. We now select some books which discuss radiation in detail, including the optical part of the electromagnetic spectrum. Two books which serve as a bridge between those on field theory and the more specialised texts on wave theory are *The theory of electromagnetism*, by D. S. Jones (Pergamon, 1964), and *Classical electromagnetic radiation*, by J. B. Marion (Academic Press, 1965). Both of these begin with Maxwell's equations and develop various aspects of the production and propagation of electromagnetic waves. For a more advanced, mathematical, treatment there is C. Müller's *Foundations of the mathematical theory of electromagnetic waves* (Springer, 1969), and *Radiation and propagation of electromagnetic waves*, by G. Tyras (Academic Press, 1969). There are many specialised monographs on the behaviour of electromagnetic waves in general, and of radiowaves in particular, including the following:

Bowman, J. J. *et al.* (eds.). *Electromagnetic and acoustic scattering by simple shapes*. North-Holland, 1969. Fifteen shapes are considered for their intrinsic importance, and as bases for synthesising the properties of more complex shapes.

Cook, A. H. *Interference of electromagnetic waves.* Clarendon, 1971. Selected topics chosen for their practical importance and their illustration of basic principles.

David, P. and Voge, J. *Propagation of waves.* Pergamon, 1969. Discusses the practical conditions of wave propagation around the earth.

Ginzburg, V. L. *The propagation of electromagnetic waves in plasmas.* 2nd edition. Pergamon, 1970.

Wait, J. R. *Electromagnetic waves in stratified media.* 2nd edition. Pergamon, 1970.

Electrodynamics is concerned with the interaction between electromagnetic fields and matter. When matter is considered in terms of atoms and particles, the study becomes quantum electrodynamics. The following are a few of the books on these two subjects:

Grandy, W. T. *Introduction to electrodynamics and radiation.* Academic Press, 1970. Deals with classical relativistic electrodynamics and non-relativistic quantum electrodynamics.

Groot, S. R. de and Suttorp, L. G. *Foundations of electrodynamics.* North-Holland, 1972. Derives the electromagnetic laws for continuous media from those for point particles.

Jackson, J. D. *Classical electrodynamics.* Wiley, 1962

Landau, L. D. and Lifschitz, E. M. *Electrodynamics of continuous media.* Pergamon, 1960. Develops the 'theory of electromagnetic fields in matter and the theory of macroscopic electric and magnetic properties of matter'.

Podolsky, B. and Kunz, K. S. *Fundamentals of electrodynamics.* Dekker, 1969. Seeks to unify classical electrodynamics by using Hamilton's principles and the symmetry properties of space and time as basic assumptions.

Källén, G. *Quantum electrodynamics.* Springer, 1972

Thirring, W. *Principles of quantum electrodynamics.* Academic Press, 1958

Static electricity

The phenomenon of static electricity has been known for a very long time and has many practical applications, but the processes involved in producing static charges are still being investigated. It also gives rise to serious hazards in some situations, and ways of overcoming unwanted electrification have to be developed. Since 1971 the Electrical Research Association have been producing *Electrostatics Abstracts*, which is published monthly and has annual indexes. It covers all aspects of the subject, from theory, hazards, applications and devices to electrostatics in chemistry and antistatics.

Among the books on the subject, *Static electrification*, by L. B. Loeb (Springer, 1958), is a useful summary of the various processes which produce static electricity, although progress has been made since it was published in understanding those processes. A more recent book is W. R. Harper's *Contact and frictional electrification* (OUP, 1967), in which the author argues that the carriers of the electric charge are never electrons when the material being charged is strictly an insulator. (Others would say that the carriers are always, or generally, electrons.) For a survey of modern electrostatics, including theory although the emphasis is more towards the practical, there is *Electrostatics and its application*, edited by A. D. More (Wiley/Interscience, 1973). A French three-volume work which covers the whole field is *Electrostatique*, by E. Durand (Masson, 1964–66).

Dielectrics

The *Digest of Literature on Dielectrics*, published annually by the US National Academy of Sciences, contains literature reviews and bibliographies ranging from fundamental properties to technological problems, including instrumentation and measurement; behaviour of liquids, solids and gases; and ferroelectric, piezoelectric and electro-optic materials. Volumes do not usually appear until over a year after the original literature was published, e.g. the *Digest* for 1971 was published in 1973.

Research papers are published in *IEEE Transactions of Electrical Insulation*, which is of interest to chemists, engineers and physicists working on dielectric materials, their properties and uses in electrical and electronic circuits and systems.

For the background theory and a discussion of materials there is *Dielectrics*, by J. C. Anderson (Chapman and Hall, 1964). A comprehensive treatment of classical dielectric theory is given in *Theory of electric polarization*, by C. J. F. Böttcher (2nd edn, Elsevier, 1973). The first volume considers dielectrics in static and low-frequency fields; the second volume will discuss their dynamic behaviour and other special topics. The following books deal with some more specialised aspects of dielectric theory and the behaviour of particular types of dielectric:

Daniel, V. V. *Dielectric relaxation*. Academic Press, 1967

Gross, B. *Charge storage in solid dielectrics, A Bibliographic review on the electret and related effects*. Elsevier, 1964

McIntosh, R. L. *Dielectric behaviour of physically adsorbed gases*. Arnold, 1966

Scaife, B. K. P. *Complex permittivity—theory and measurement*. English Universities Press, 1971

Zheludev, I. S. *Physics of crystalline dielectrics*. 2 vols. Plenum, 1971

Two books concerned with the measurement of properties are *Electrical insulation measurements*, by W. P. Baker (Newnes, 1965), intended for industrial laboratory workers and supply engineers; and *High frequency dielectric measurement*, edited by J. Chamberlain and G. W. Chantry (IPC Science and Technology Press, 1973), which contains the proceedings of a tutorial conference held at the National Physical Laboratory.

Monographs on the electrical breakdown of dielectric liquids and solids include *Ionization, conductivity and breakdown in dielectric liquids*, by I. Adamczewski (Taylor and Francis, 1969), and *The theory of dielectric breakdown of solids*, by J. J. O'Dwyer (Clarendon, 1964).

Electric discharges in gases

Having just noted some books on discharges in dielectric liquids and solids, it will be convenient to outline some of the literature on the ionisation and breakdown of gases. Professor F. Llewellyn Jones is the author of three introductory monographs on this subject: *Ionization and breakdown in gases* (Methuen, 1957); *Ionization avalanches and breakdown* (Methuen, 1967), which is essentially additional chapters to the previous book and includes breakdown under extreme conditions; and *The glow discharge* (Methuen, 1966), which discusses the physics of the phenomenon and outlines its applications in electronics and plasma physics. Two further texts intended for the research worker are *Ionized gases*, by A. Von Engel (2nd edn, Clarendon Press, 1965), and *Electron avalanches and breakdown in gases*, by H. Raether (Butterworths, 1964). The latter includes a review of methods of observing avalanches. A more specialised book on experimental and theoretical aspects of the interaction between microwaves and gases is A. D. MacDonald's *Microwave breakdown in gases* (Wiley, 1966).

Resistance and resistors

Many of the works on this aspect of electricity have been written for the electrical engineer and are mainly concerned with the different types of resistor. C. L. Wellard's *Resistance and resistors* (McGraw-Hill, 1960) is also intended for engineers, but covers the basic facts of resistance in addition to details of the more common types of resistor. The modern theory of electrical resistance of metals and alloys is presented in *Electrical resistance of metals*, by G. T. Meaden (Plenum, 1965), and also useful in this study is *Electric conduction in semiconductors and metals*, by W. Ehrenberg (OUP, 1958).

Semiconductors and transistors

The semiconductor, so called because it has an electric conductivity between that of a metal and an insulator, has been the cause of a minor literature explosion in recent years, including considerable contributions from solid state physicists on the one hand and electronic engineers on the other. Here it is possible only to indicate some of the research journals and give a selection of monographs to guide anyone beginning to work on the subject. Two of the journals which specialise in this field are the monthly *Solid-state Electronics* (Pergamon), and *Soviet Physics—Semiconductors* (AIP), which is a cover-to-cover translation of *Physics and Technics of Semiconductors of the Academy of Sciences of the USSR*. Both of these include theory and applications. A regular conference series is the *International Conference on the Physics of Semiconductors* sponsored by the International Union of Pure and Applied Physics. These conferences are held every 2 years, and the publisher of the proceedings varies, depending on the country in which the conference was held. A book series in which each volume deals with a particular topic is *Semiconductors and semimetals*, edited by R. K. Willardson and A. C. Beer (Academic Press, 1966–).

A useful brief introduction to the properties of semiconductors is given in D. A. Wright's *Semi-conductors* (Methuen, 1966). Another introductory text which summarises the theory and includes information on the design and operation of devices is *Semiconductors*, by H. F. Wolf (Wiley/Interscience, 1971). The optical and thermal properties of semiconductors have important applications in energy conversion and are discussed in *Photo- and thermo-electric effects in semiconductors,* by S. M. Ryvkin (Consultants Bureau, 1964), and

Semiconductor opto-electronics, by T. S. Moss *et al.* (Butterworths, 1973).

In addition to the familiar transistor there are several important semiconductor devices such as the thermistor and the photoelectric cell. *Physics and technology of semiconductor devices*, by A. S. Grove (Wiley, 1967), discusses these devices in general. Similar coverage is found in *Physics of semiconductor devices*, by S. M. Sze (Wiley/ Interscience, 1969). The circuitry involved is included in *Handbook of semiconductor electronics*, by L. P. Hunter (3rd edn., McGraw-Hill, 1970). The books on transistors are legion. Introductory books on their theory and applications include *Transistors—theory and circuitry*, by K. J. Dean (McGraw-Hill, 1964); *Transistor physics*, by K. G. Nichols and E. V. Vernon (Chapman and Hall, 1966); and *Introduction to transistor electronics*, by R. L. Walker (Blackie, 1966).

Superconductivity

When the temperature of certain materials falls to near absolute zero, their electric resistivity vanishes. The literature of low-temperature physics in general has been considered in the chapter on Heat and Thermodynamics, so here we shall concentrate only on the literature of this electrical property.

A fairly full treatment, intended for students, is to be found in *Introduction to superconductivity*, by A. C. Rose-Innes and E. H. Rhoderick (Pergamon, 1969), while *Superconductivity*, edited by R. D. Parkes (2 vols., Dekker, 1969), is a comprehensive treatise covering all aspects of the phenomenon.

A successful phenomenological theory for superconductivity was developed by Fritz and Heinz London in 1935, and this is described in *Superfluids, Volume 1: Macroscopic theory of superconductivity*, by F. London (2nd edn, Dover, 1961). When this book first appeared in 1950, the modern period in the development of superconductivity theory began through the discovery by Fröhlich of the electron–phonon interaction. The subsequent advances in the microscopic theory are explained in J. M. Blatt's *Theory of superconductivity* (Academic Press, 1964). The pairing theory, known as the BCS theory after its orginators, Bardeen, Cooper and Schrieffer, is presented in *Theory of superconductivity*, by J. R. Schrieffer (Benjamin, 1964), and *Theory of superconductivity*, by G. Rickayzen (Interscience, 1965).

Superconductivity has found many useful applications, which are outlined in books such as *Applied superconductivity*, by V. L. Newhouse (Wiley, 1964), and *Superconductivity and its applications*, by J. E. C. Williams (Pion, 1970). Recent advances in the uses of super-conductivity are to be found in the proceedings of the *Applied Superconductivity Conference*. These were published at one time in the *Journal of Applied Physics*, but the 1972 conference proceedings were published separately by the Institute of Electrical and Electronics Engineers.

Piezo- and ferroelectricity

When mechanical pressure is applied to certain dielectric crystals, known as piezoelectrics, an electric polarity is produced which will give rise to a flow of electricity if a circuit is completed. Ferroelectrics are substances which have a permanent, spontaneous electric polarisation, and all of them exhibit piezoelectricity. A journal which published papers on both these phenomena, and on related topics such as electro-optics, pyroelectrics, non-linear dielectrics and liquid crystals, is *Ferroelectrics* (Gordon and Breach). The following is a representative selection of books on these related topics:

Burfoot, J. C. *Ferroelectrics: an introduction to the physical principles.* Van Nostrand, 1967

Cady, W. G. *Piezoelectricity: an introduction to the theory and applications of electromechanical phenomena in crystals.* 2 vols. Dover, 1964

Fatuzzo, E. and Merz, W. J. *Ferroelectricity*. North-Holland, 1967. On the basic physics of the subject.

Grindlay, J. *An introduction to the phenomenological theory of ferroelectricity*. Pergamon, 1970.

Energy conversion

The quarterly journal with the title *Energy Conversion* (Pergamon) carries articles on all methods of direct and dynamic conversion with their regulation and control.

Among sources of electricity there are primary and secondary batteries, and the various methods of direct energy conversion (fuel cells, photoelectricity, thermoelectricity, magnetohydrodynamic power generation, etc.). The latter methods have attracted consider-able attention recently and much has been written on them. Once again a selection of books has had to be made, concentrating as far as possible on those which emphasise physical principles.

Batteries

Heise, G. W. and Cahoon, N. C. (eds.). *The Primary battery*. Wiley Vol. 1, 1971. An authoritative and comprehensive text.

Jasinski, R. *High-energy batteries*. Plenum, 1967. A general introduction.

Falk, S. U. and Salkind, A. J. *Alkaline storage batteries*. Wiley, 1969. A comprehensive reference book.

Direct energy conversion—general

Angrist, S. W. *Direct energy conversion*. 2nd edition. Allyn, 1971. Includes physics, and a discussion of the state of the art for devices.

Spring, K. H. (ed.). *Direct generation of electricity*. Academic Press, 1965. Concentrates on physical principles.

Thermoelectricity

Barnard, R. D. *Thermoelectricity in metals and alloys*. Taylor and Francis, 1972. Account of recent understanding of the phenomena for both liquid and solid state.

Harman, T. C. and Honig, J. M. *Thermoelectric and thermomagnetic effects and applications*. McGraw-Hill, 1967. An advanced treatment of selected topics with special reference to energy conversion processes.

Macdonald, D. K. C. *Thermoelectricity: an introduction to the principles*. Wiley, 1962. A short introduction for experimental physicists.

Solar cells

Ranney, M. W. *Solar cells* (Noyes Development Corp., 1969). Based on US Patents relating to solar cells.

Solar cells. Proceedings of the International Colloquium Organized by the European Cooperation Space Environment Committee for July 6 to 10, 1970 in Toulouse, France. Gordon and Breach, 1971

Electrical measurements

Many of the books in the list that follows have been written with the requirements of the electrical engineer in mind, but they do provide the physicist also with information on the principles and uses of the various instruments and methods which are available.

Bleuler, E. and Haxby, R. O. (eds.). *Electronic methods.* Academic Press, 1964. Written for the research worker who must design his own equipment.

Golding, E. W. and Widdis, F. C. *Electrical measurements and measuring instruments.* 5th edition. Pitman, 1963.

Hague, B. *Alternating current bridge methods.* 6th edition. Pitman, 1971. A standard reference book.

Karsa, B. E. F. *Electrical measuring instruments and measurements.* Akadémiai Kiadó, Budapest, 1967. A comprehensive text on principles, instruments and methods.

Schwab, A. J. *High-voltage measurement techniques.* MIT, 1972

Electronics

It is obvious that a comprehensive description of the literature of electronics is impossible here, but physicists are often concerned with the subject, from either a theoretical or a practical point of view, so that a brief outline of some of the sources of information is necessary.

There are several reference books, including *A dictionary of electronics*, by S.Handel (2nd edn, Penguin, 1971); *Dictionary of electronics and nucleonics*, edited by L. E. C. Hughes *et al.* (Chambers, 1969); and *Electronics and nucleonics dictionary*, by J. Markus (3rd edn, McGraw-Hill, 1966). The last two cover nuclear science and engineering as well as electronics. Another valuable reference work, which is intended for the practising engineer is the *Electronic engineers' reference book*, edited by L. E. C. Hughes and F. W. Holland (3rd edn., Heywood, 1967); it covers all aspects, including the general physical background, properties of the ionosphere, radiation, components, computing, etc.

Most of the abstracts journals mentioned at the beginning of this section include electronics in their subject coverage. There are several others devoted purely to electronics and related topics including *Electronics and Communications Abstracts* (Multi-science Publishers) and, for current Russian literature, *Electronics Express* (In. Physical Index).

An important guide to the literature of electronics is *Electronics: a bibliographical guide*, by C. K. Moore and K. J. Spencer (Macdonald, 1961). The first volume covering the years 1945–59 was published in 1961; the second volume covering the period July 1959 to December 1964 was published in 1965; a third volume is planned. The guide is divided into a large number of sections beginning with reference works (abstracting, bibliographical and translation services) and then going on to the specialist subject fields. Each section contains a list of all relevant bibliographies with a selection of authoritative books and most important papers (pioneer papers, surveys, special journal issues, papers with important lists of references, etc.). References include full bibliographical details, and many are briefly annotated. Another bibliographical tool is edited by the staff of the Electronic Properties Information Center, Hughes Aircraft Co., Culver City, California, and is published under the title *Electronic properties of materials: a guide to the literature* (IFI, Plenum). Volume 1 appeared in 1965, Volume 2 in 1967 and Volume 3 in 1971. Each volume is in two parts—the first part gives a list of materials and properties which is linked by reference numbers to the second part, where bibliographical references are given. The first volume contains a thesaurus for materials and properties as a guide to the terms which are actually used.

There is a very wide range of journals devoted to one or more aspects of electronics. Many of those listed at the beginning of this section on electricity include articles on electronics. Of the others the *International Journal of Electronics*, published monthly by Taylor and Francis, covers both theoretical and experimental work; *Radio Engineering and Electronic Physics* (Scripta), which is a translation of the Russian serial *Radiotekhnika i Elektronika,* includes a significant amount of fundamental physics; and similarly for Japanese research there is the journal *Electronics and Communications in Japan* (Scripta), which is a translation of *Denshi Tsushin Gakkai Ronbunshi.*

As far as textbooks are concerned, it is possible here only to select a few useful introductory books on electronics and electronic circuits. The *Bibliographical guide* mentioned above will provide a comprehensive list of books on the whole of electronics:

Chirlion, P. M. *Electronic circuits—physical principles, analysis and design.* McGraw-Hill, 1971.

Owen, G. E. and Keaton, P. W. *Fundamentals of electronics.* 3 vols. Harper and Row, 1966, 1967. The first volume is concerned with mathematical methods for circuit analysis; the second mainly with vacuum tubes and semiconductors; and the third discusses various circuits.

Rollin, B. V. *An introduction to electronics.* Clarendon, 1964. Intended for physics and engineering students.

Ryder, J. D. *Electronic fundamentals and applications.* 4th edition. Pitman, 1970

Electron optics

We close this section on the literature of electricity by looking at some of the books which discuss the optical properties of electron beams and their application in electron microscopy. A comprehensive work covering both optics and instruments is P. Grivet's *Electron optics* (2nd edn, Pergamon, 1972). *Photoelectric imaging devices*, edited by L. M. Biberman and S. Nudelman (Plenum, 1971), is an encyclopaedic work in two volumes—one concerned with physical processes and methods of analysis, and the other with devices and their evaluation. A concise account of electron diffraction including apparatus, theory and applications is found in *Electron diffraction*, by T. B. Rymer (Methuen, 1970).

There are several types of electron microscopy, which are described in books such as *Field ion microscopy*, by E. W. Müller and T. T. Tsong (American Elsevier, 1969); *Fundamentals of transmission electron microscopy*, by R. D. Heidenreich (Interscience, 1964); and *The scanning electron microscope*, by C. W. Oatley (CUP, Vol. 1, 1972).

MAGNETISM

Because of the intimate relationship between electricity and magnetism, several of the books and journals already mentioned under Electricity cover both subjects. This is particularly true of the abstracts serials and the introductory textbooks, and the following paragraphs, therefore, to some extent supplement what has already been written.

Abstracts and general journals

The Bell Telephone Laboratories produce an abstracts journal covering all parts of magnetism, with the title *Index to the Literature of Magnetism*. A literature survey, *Magnetism and Magnetic Materials Digest*, is published annually by Academic Press; each volume reviews the previous year's literature. The two principal journals on magnetism are the *International Journal of Magnetism* (Gordon and Breach) and the *IEEE Transactions on Magnetics*. Both of these cover the whole of magnetism from both theoretical and applied viewpoints. From time to time the IEEE journal consists of a special issue with the title *Advances in Magnetics*, containing literature surveys and critical reviews.

Textbooks and conferences

Student texts have been noted in the section on electricity since most of them include chapters on magnetism, but one which concentrates solely on magnetism is *Introduction to the theory of magnetism*, by D. Wagner (Pergamon, 1972). It is divided into three sections—diamagnetism, paramagnetism and ferromagnetism. Also intended for university students, but including a great deal of experimental work not normally found in textbooks, is *Modern magnetism*, by L. F. Bates (4th edn, CUP, 1961). *Magnetism in solids*, by D. H. Martin (Iliffe, 1967), has been written for recent graduates or for those who need to be brought up to date with newer developments. An older book which has been reprinted several times and retains much of its value as an exposition of both classical and quantum theories of magnetism is *The theory of electric and magnetic suscepti-bilities*, by J. H. Van Vleck (OUP, 1932; reprinted 1965). A more recent book on the quantum theory is R. M. White's *Quantum theory of magnetism* (McGraw-Hill, 1970). *The physical principles of magnetism*, by A. H. Morrish (Wiley, 1965), is for graduates with a knowledge of solid state physics and quantum theory. A major, multivolume, work on modern magnetism is *Magnetism: a treatise on modern theory and materials*, edited by G. T. Rado and H. Suhl (Academic Press, 1963–66).

Magnetic domain theory has been receiving an increased amount of attention in recent years. An introductory text has been written by R. S. Tebble—*Magnetic domains* (Methuen, 1969)—while a considerable amount of experimental detail has been included in *Magnetic domains and techniques for their observation*, by R. Carey and E. D. Isaac (English Universities Press, 1966).

The *Proceedings of the International Conference of Magnetism*, held in Nottingham in 1964 and published in the following year by the Institute of Physics and the Physical Society (London), contains over 200 papers, concerned mainly with theory and magnetic materials. An annual Conference on Magnetism and Magnetic Materials is held in the USA. Up to 1970 the proceedings were published in the *Journal of Applied Physics*. From 1971 onwards they have been published in the *Conference Proceedings Series* of the American Institute of Physics.

Magnetic materials—theory and properties

A short introduction which is mainly devoted to Ferromagnetics is

Magnetic materials, by F. Brailsford (3rd edn, Methuen, 1960). *Structure and properties of magnetic materials*, by D. J. Craik (Pion, 1970), is another introductory text; the first few chapters discusses general magnetism, and atomic and crystal structure; the rest of the book is about magnetic behaviour of materials, with an extensive treatment of domain structures. The magnetic properties of a wide range of materials are presented in *Magnetic materials*, by R. S. Tebble and D. J. Craik (Wiley/Interscience, 1969). Another reference book which incorporates a great deal of information on the applications of magnetic materials in various electrical and electronic devices is F. N. Bradley's *Materials for magnetic functions* (Heyden, 1971). A specialised text concerned with particular types of magnetic materials is *Rare earth permanent magnets*, by E. A. Nesbitt and J. H. Wernick (Academic Press, 1973).

Ferromagnetism

This is the most familiar type of magnetism, in which magnetic bodies have a strong attraction for one another. There is a comprehensive German text on the subject by E. Kneller, *Ferromagnetismus* (Springer, 1962), and a rather older, but still useful, book in English— *Ferromagnetism*, by R. M. Bozorth (Van Nostrand, 1951). The domain structure of ferromagnetics and a summary of the related experimental work is given in *Ferromagnetism and ferromagnetic domains*, by D. J. Craik and R. S. Tebble (North-Holland, 1965). *Micromagnetics*, by W. F. Brown (Interscience, 1963), analyses ferromagnetic materials on a scale intermediate between domains and individual atoms.

Ferrimagnetism

This is a type of permanent magnetism exhibited by materials known as ferrites which have come to play an essential role in electronics, as core materials, in magnetic recording heads and in microwave components. *Oxide magnetic materials*, by K. J. Standley (2nd edn, Clarendon, 1972), presents an elementary account of ferrimagnetism and the uses of ferrites and garnets. Papers covering a wide range of research on ferrites and their uses are to be found in *Ferrites— Proceedings of the International Conference*, edited by Y. Hoshino *et al.* (University of Tokyo Press, 1971). Those ferrites which are magnetically soft and are used in inductors, transformers and related devices are discussed in *Soft ferrites: properties and applications*, by E. C. Snelling (Iliffe, 1969). There are numerous books on the microwave applications of ferrites. *Handbook of microwave ferrite materials*, edited by W. H. von Autlock (Academic Press, 1965), reviews work on these materials published from 1950 to 1963 and includes hundreds of tables and graphs of their properties. A useful book on this application of ferrites is *Principles of microwave ferrite engineering*, by J. Helszajn (Wiley/Interscience, 1969).

Magnets and magnetic fields—production, measurement and uses

A general text on the applied side of magnetism is *Experimental methods in magnetism*, by H. Zijlstra (2 vols., North-Holland, 1967); the first volume discusses the generation and computation of magnetic fields, and the second the measurement of magnetic quantities.

On the subject of permanent magnets and their uses there is *Dauermagnete: Werkstoffe und Anwendungen*, by K. Schüler and K. Brinkmann (Springer, 1970); and two somewhat older books in English: *Permanent magnets and magnetism*, edited by D. Hadfield (Iliffe, 1962), and *Permanent magnets and their applications*, by R. K. Parker and R. J. Studders (Wiley, 1962).

The design of electromagnets is the chief subject of *Laboratory magnets*, by D. J. Kroon (Philips Technical Library, 1968). Also design-oriented is a book which aims at being a complete reference to all types of solenoid magnet, *Solenoid magnet design: the magnetic and mechanical aspects of resistive and superconducting systems*, by D. B. Montgomery (Wiley/Interscience, 1969).

Two further books which are concerned with the production of these high fields are *The generation of high magnetic fields*, by D. H. Parkinson and B. E. Mulhall (Heywood, 1967); and *Pulsed high magnetic fields*, by H. Knoepfel (North-Holland, 1970), which deals with transient fields lasting for less than 0·1 s, and includes an extensive bibliography.

13

Nuclear and atomic physics

A. P. Banford and *E. Marsh*

The term 'Nuclear and Atomic Physics' is here taken to mean the study of the structure and properties of atomic nuclei, and of sub-nuclear or elementary particles; the latter branch of the subject is also known as high-energy physics. The physics of atoms and ions —that is to say, the phenomena involving only the orbital electrons— is not covered. In the supporting technologies a dividing line is drawn so as to include instrumentation for nuclear physics and particle accelerators, but excluding reactor technology and applications of radioactive isotopes.

The subject has enjoyed support on a substantial scale since the war, largely on account of the utilisation potential of nuclear energy for both peaceful and military purposes. Nuclear physics is also characterised by the large dimensions and complexity of the equipment and by a rather luxuriant literature. There are many specialist journals, but a substantial number of papers are published in general journals, particularly if these are prestigious or can publish more rapidly. A common effect of the scale of research in experimental high-energy physics is multiple authorship in an extreme form. Fifty authors to a paper is not unknown, and papers are often published by 'collaborations'—no individuals being named, only the institutions.

In elementary particle physics, the preprint, with all its attendant bibliographical problems, is a well-established medium of communication, as is the conference. In this field progress can be rapid (despite the fact than an experiment can take 3 years), and researchers maintain a variety of informal communication networks so as to keep fully up to date. English is the lingua franca of nuclear physics, and the vast majority of papers are published in that language even when it is not the first language of author or journal or either. The most important Russian journals are available in cover-to-cover English translation.

PRIMARY JOURNALS

The leading international specialist journal is *Nuclear Physics*, launched in 1957 by North-Holland and covering both theory and experiment. After 10 years, binary fission occurred; *Nuclear Physics B* covering particle physics was instituted and *Nuclear Physics A*, which continued the original volume numbering sequence, thereafter confined itself to the medium-energy phenomena involving the nucleus as a whole. *NPA* is now in weekly issues of about 200 pp., making some 20 volumes per year. *NPB* (with separate pagination) is published fortnightly, and currently amounts to 15 volumes annually. *A* and *B* share a common subject indexing *system* (though the indexes themselves are issued separately). The major categories include Nuclear Structure, Nuclear Reactions, Electromagnetic Radiation, Quantum Field Theory, Fundamental Interactions, with a detailed subdivision that is markedly different from those used by INSPEC and the American Institute of Physics. Since 1965 authors of experimental papers on low-energy nuclear physics have themselves been required to select, from a published list, keywords or symbols to help index their papers; the keywords are printed below the abstract. The intended extension to the other topics covered by this journal has not materialised. Both parts of *Nuclear Physics* are currently publishing papers on average 4–5 months after receipt, which is the stated intention of the editors. The journal does not offer a rapid publication service for letter-type communications, nor is there (as in the case of *Physical Review and Nuovo Cimento*) an associated letter journal.

North-Holland also publish *Nuclear Instruments and Methods*, which covers not only new particle detectors, accelerator techniques and electronic circuits, but also mathematical aspects such as the treatment of observational data, the analysis of experimental geometries and particle optics. Issues are fortnightly, with three issues making a volume; papers appear 4–5 months after receipt, letters somewhat more quickly. The 'subject' indexing of this journal is very poor. There are only eight categories with no sub-division whatsoever; one category (Instruments and Methods for Nuclear Spectroscopy) accounts for over 40% of articles. Letters are listed separately, with no subject classification at all. As in the case of *Nuclear Physics*, cumulative indexes (subject and author) are issued every 10 volumes, but the shallowness of the subject indexing in *NIM* makes them much less useful.

Papers on the basic physics, design concepts and instrumentation of particle accelerators now tend to appear in the quarterly journal *Particle Accelerators* (Gordon and Breach), which appeared in 1970. Subject indexing is at one level only, but, with 18 categories and fewer papers per volume than *NIM*, is acceptable. Reactor physics and associated techniques are covered by the American Nuclear Society's monthly *Nuclear Science and Engineering*. Here again the 'subject' index could be better. Articles are listed alphabetically by contracted title, with a certain amount of rotation to give essentially a KWIC index.

Cover-to-cover translations of Russian journals include *Soviet Journal of Nuclear Physics* (*Yadernaya Fizika*), *Soviet Atomic Energy* (*Atomnaya Energiya*) and *Soviet Journal of Particles and Nuclei* (*Fizika Elementarnykh Chastits i Atomnogo Yadra*); the last-named, known as *Particles and Nuclei* for the first two volumes, has been retitled to avoid confusion with the longer-established American journal of the same title published by the Fazl-I-Umar Research Institute of Athens, Ohio. *SAE* is devoted to applied nuclear physics, reactors and accelerators. The Russian original of *SJPN* was introduced in 1970 by the Joint Institute for Nuclear Research, Dubna. Papers on high-energy physics still appear in *SJNP* (USSR Academy of Sciences) in addition to those on lower-energy nuclear physics. There is a delay of nominally 6 months between the appearance of the Russian original and its translation; coupled with the sometimes substantial delay between submission and original publication, this means that much Russian work only becomes known in the West after a long time (unless, as often occurs, preprints are distributed). None of these three journals has a subject index and only *SJNP* has an author index.

Despite the existence of these specialist periodicals, a large amount of original research is published in other journals—some of which, it is true, have split into nuclear and non-nuclear parts. Rightly or wrongly, many authors regard the *Physical Review* or *Physical Review Letters* as the best place to publish, even though page charges and restrictions on length are involved. An extremist fringe tends to regard all other journals as obscure and unlikely to be read by the right people, and they look upon a paper in *PR* or *PRL* as possessing some special extra quality.

Physical Review has undergone several changes of format in order to keep individual issues of the journal to manageable size. It is currently divided into four parts (*A*, *B*, *C* and *D*), with separate pagination for each. Part *C* is devoted to Nuclear Physics and appears once a month, while part *D*, which deals with Particles and Fields, is published twice a month. Each part has two volumes per year. The average delay between receipt of a paper and its publication is 6 months.

The American Physical Society also publishes the companion *Physical Review Letters*, a weekly journal 'containing short communications dealing with important new discoveries or topics of high current interest in rapidly changing fields of research'. The editors adhere rigorously to these criteria in order to preserve the essential character of *PRL*. For instance, contributions must be less than three-and-a-half pages, and from time to time fields of research which are no longer rapidly changing are weeded out. Letters are published on average 15 weeks after submission, though delays as short as 5 weeks are on record. One single index covering a volume of *PR* (all four parts) *and* a volume of *PRL* is issued every 6 months. The subject index for nuclear physics is divided first by atomic number, then by property or reaction type; for particle physics the separation is primarily into theory and experiment, then according to the interaction involved (i.e. weak, electromagnetic or strong). There is no classification by the type of particle, which, in a sense, is inconsistent with the nuclear physics primary sub-division by size of nucleus.

Physics Letters (North-Holland) is now in two parts, with nuclear and particle physics in Part B. Publication is weekly (Parts *A* and *B* alternating), with typically a 2-month delay between receipt and publication of a letter. No subject indexes are published.

Journal of Physics, published by the (British) Institute of Physics, was for many years not popular with authors of papers on nuclear and particle physics. But now Part A (monthly) is sub-titled 'Mathematical, Nuclear and General' and each issue carries several papers on elementary particle and nuclear physics, and cosmic-rays. The subject indexing is very similar to that in *Physics Abstracts*; there are many 'see' and 'see also' references and there is a helpful explanation of the structure of the index.

Nuovo Cimento (Società Italiana di Fisica) attracts papers on particles and on field theory, which appear in Part *A*: papers on nuclear physics are comparatively rare. Almost all the papers are in English. The subject indexing is poor. The associated *Lettere al Nuovo Cimento*, introduced in 1969, is well patronised by those seeking rapid publication. *Zeitschrift für Physik* (Springer) often carries papers on nuclear and particle physics; those in German have an English abstract. The Japanese journal *Progress of Theoretical Physics*, in which all papers are in English, attracts contributions on (*inter alia*) theoretical nuclear and particle physics mainly, but not exclusively, from Japanese authors. *Instruments and Experimental Techniques* (a cover-to-cover translation of the Russian *Pribori i*

Tekhnika Eksperimenta) has many articles on particle detector and accelerator techniques; such papers are also occasionally to be found in *Review of Scientific Instruments* and *Journal of Physics E* (formerly *Journal of Scientific Instruments*).

REVIEW JOURNALS

Nuclear and particle physics has proved to be a fertile breeding ground for the 'Advances in . . .' type of publication. Not all the seedlings have survived, despite the fact that in such a rapidly advancing subject there is a real need for up-to-date review articles.

Without doubt the leading position is occupied by *Annual Review of Nuclear Science* (Annual Reviews Inc.). Each volume contains about 10 articles, typically of 40–50 pages and with substantial bibliographies, contributed by acknowledged leaders in their fields. The topics which are covered are particle physics, cosmic rays, nuclear physics and chemistry, reactors, accelerators and particle detectors. Since each volume contains a cumulative subject index to the previous 10 volumes, there is no need to list specific articles here, though mention must be made of H. A. Bethe's article on the 'Theory of nuclear matter' (*21*, 93–244, 1971), with 211 references.

Progress in Nuclear Physics (Pergamon) appeared every year or two until 1963 (vol. 9) when there was a 6-year gap. Then two further volumes appeared, followed by two parts of vol. 12. Since late 1970 nothing has been issued, and the future of this publication is uncertain. Both theoretical and experimental nuclear and particle physics is covered, together with the associated instrumentation. Volume 11 has a cumulative contents list. *Progress in Nuclear Techniques and Instrumentation* (North-Holland) had a brief existence of three volumes between 1965 and 1968. During this period 13 review articles on detectors, accelerators and similar topics were published, but most have been superseded by later articles published elsewhere, or rendered obsolete by later technical advances.

Advances in Nuclear Physics (Plenum) is currently thriving, and tends to carry articles on more specialised topics than does *ARNS*. *Advances in particle physics* (Interscience) appeared for Vol. 1 and 2 in 1968 and had not been heard of since.

The above review publications are all in book format, appearing roughly annually. This results sometimes in an excessive period of time between the latest reference cited and the publication of the review itself. Gordon and Breach publish two journals at relatively frequent intervals. *Fields and Quanta* comes out quarterly and carries review articles of the conventional types without necessarily including exhaustive bibliographies. One recent issue (3 July 1972) was entirely devoted to a review by Y. S. Kim and N. Kwalk on the current status of the quest for quarks (138 pp., 108 references, the latest dated March 1971). *Comments on Nuclear and Particle Physics* (every 2 months) is unusual in that the articles are contributed only by a panel of about 30 correspondents, who write short articles (typically five pages, commenting critically on recent developments. The papers do not pretend to be definitive statements of the current state of the art, but are intended to stimulate further thought by posing new questions or by offering a new viewpoint in recent work.

Atomic Energy Review is published by the International Atomic Energy Agency, in quarterly parts. It carries extensive, specially commissioned, articles on nuclear technology, data compilations and other useful topics, together with reports on conferences. Articles are in English, with French, Russian and Spanish abstracts.

Reviews on nuclear and particle physics may also be found in non-specialist publications. *Reviews of Modern Physics* publishes the review of particle properties in its April issue, and also carries conventional reviews in this field. *Rivista al Nuovo Cimento* and *Physics Reports* (*Physics Letters C*) have review articles in this field to an extent that mirrors the coverage of the parent journals. No one, not even the active specialist, should ignore the occasional article in *Scientific American*. The authors are persons of standing, and the clarity of the writing and the quality of the illustrations are usually first rate.

CURRENT-AWARENESS PUBLICATIONS

It is quite common for the conventional abstract journal to be scanned regularly for current-awareness purposes as well as being used for retrospective searches; these are dealt with later. Here we are concerned with those services which are concerned primarily with rapid notification, if necessary at the cost of elegance and an abstract.

Particularly in the field of particle physics, the preprint (see below) is a popular medium, and several organisations issue lists of preprints received. The Stanford University Linear Accelerator Center issues *Preprints in Particles and Fields* weekly, with a production delay of only a few days achieved by computer listing methods. Each issue (usually a single A4 sheet) lists about 70 preprints (title, author, code number plus an indication of whether the preprint is experimental or theoretical). From time to time anti-preprint lists are issued, giving open literature references for preprints that have been published. The coverage is confined to physics only, not the supporting techniques.

The *International Nuclear Information System* (*INIS*) was launched by the IAEA in 1970. It is a computer-based system 'for identifying published information about nuclear science and its peaceful applications'. Input is provided by atomic energy organisations in the member states, who select items for inclusion and assign descriptors from the *INIS Thesaurus*. Four outputs are available: magnetic tapes (available only to participating organisations and usable for national current-awareness services by computer searching), the *INIS Atomindex*, abstracts on microfiche and full text on microfiche. The last-named is confined to non-conventional literature (i.e. anything other than books and journals), and individual items may be ordered.

The subject scope was initially confined to applied nuclear science, but from 1972 there has been full coverage of the subject. The subject categories are Physical Sciences (General Physics, High-Energy Physics, Neutron and Nuclear Physics), Chemistry, Materials and Earth Sciences, Life Sciences, Isotopes, Engineering and Technology, Other Aspects. Each category is subdivided twice. The first-level sub-divisions for Physical Sciences are shown in brackets above. High-Energy Physics is further sub-divided into Theory and Experiment, while Neutron and Nuclear Physics has four sub-divisions, including Nuclear Theory and Nuclear Properties and Reactions. The techniques of nucler physics are covered by the other categories; Radiation Protection comes under Life Sciences, Radiation Sources under Isotopes, while the Engineering and Technology category includes Accelerators, Reactors and Instrumentation. The *INIS Atomindex* is now issued twice a month, each issue listing 4000–5000 items. They are grouped under the subject categories mentioned above. For each item, in addition to the normal bibliographical details, there is listed the descriptors that were assigned to the item at the input stage. Although abstracts are also prepared, they are not reproduced in *Atomindex*, being available only by subscription (not individually) on microfiche. Each issue has author and report number indices, cumulated each May and December. No descriptor index is published, although presumably this could be done easily enough, since all the descriptors are on the magnetic tapes, so that producing such an index is a purely mechanical task; just such an index is incorporated in the *DESY High Energy Physics Index*, also computer-produced (see below). For current-awareness purposes, the lack of such an index is no hardship, but it means that retrospective searches can be done only on the (not generally available) magnetic tapes or by laboriously examining the appropriate section of each individual issue.

Mention can also be made here of the monthly *CERN Courier*, which started life as that laboratory's house journal but which has now broadened its scope to include reports on advances in particle physics, accelerator technology and similar matters from similar laboratories throughout the world. English and French editions are published.

The subject is, of course, covered by the general current-awareness publications such as *Current Papers in Physics* (INSPEC).

The *High Energy Physics Index*, compiled by DESY (Hamburg) and published by ZAED, appears every 2 weeks and deals with publications on experiments, theory and instrumentation in elementary particle physics. About 10 000 items per year are listed from all forms of publications. For preprints and similar material the service relies (as does *PPF*) on authors sending in their work. No abstracts are published, but each entry has keywords assigned to it. These keywords are used in compiling the subject index (prepared for each issue), and for an associated computerised information service. A thesaurus of keywords is issued. They are used in combination with each other or with non-keywords. A small sample test (in experimental high-energy physics) shows that *HEPI* lists items 6 weeks before *NSA*, and that *HEPI* includes more preprints and internal reports. *NSA*, however, has the better record with cumulative indexes.

ABSTRACTS JOURNALS

The subject of nuclear and particle physics is well served with abstracts journals, among which *Nuclear Science Abstracts* is outstanding. It was instituted in 1948, and is published by the Technical Information Center of the US Atomic Energy Commission. The coverage is about 60 000 items per year. Parts are issued twice a month. From 1973 there are two volumes per year, with indexes cumulated for each volume. This replaces earlier arrangements whereby indexes were cumulated quarterly for each yearly volume and then for 5-year periods. A very attractive feature of this journal is the provision of a full subject index (as well as author and other indexes), in each half-monthly part, which together make up half the physical bulk of each issue.

The coverage of *NSA* is very wide, as regards both subject matter and source material. The first issue of each volume carries a detailed description of the indexing methods and journal coverage (see also TID 4579). Within each issue, abstracts are arranged in sections with one- or two-level sub-division. The section headings, which give a good idea of the subject coverage, are: Chemistry, Engineering, Environmental and Earth Sciences, Instrumentation, Isotope and Radiation Source Technology, Life Sciences, Materials, Nuclear Materials and Waste Management, Particle Accelerators, Physics (Astrophysics and Cosmology), Physics (Atmospheric), Physics (Atomic and Molecular), Physics (Electrofluid and Magnetofluid), Physics (High Energy), Physics (Low Temperature), Physics (Nuclear), Physics (Plasma and Thermonuclear), Physics (Radiation and Shielding), Physics (Solid State), Physics (Theoretical), Reactor Technology, General.

In addition to the obvious coverage of the USAEC's own publications, *NSA* also includes journal, report, book, patent and conference literature on a world-wide basis. This comprehensive coverage is made possible partly by the scanning and abstracting done by 10 national atomic energy agencies of their own country's output. Well over 2000 journals are scanned, report coverage is unrivalled, and preprints and conference papers are included in advance of their formal publication. Numerous publishers provide *NSA* with advance page-proof copy to help minimise the delays involved, which currently average 3 months for articles in the commoner journals.

Subject indexing is done by grouping entries under headings chosen from a controlled list, details of which are given in the report TID 5001. Headings are normally names of materials, devices, concepts, theories or processes. TID 5001 is liberally provided with cross-references to related or more specific headings (as is the annual cumulative subject index, but not that in each issue). Within a subject heading, entries are ordered alphabetically by moderator words (free language but with certain preferred and non-preferred terms) which, together with the heading, constitute a string of keywords for the item. TID 5001 is revised every year or two. A consequence of this is that a topic may 'disappear' during a retrospective search that goes back to the time when the now standard terminology did not exist. This is a minor disadvantage to suffer for the benefits of an indexing framework that keeps in touch with the development of the subject.

From the French Atomic Energy Commission (CEN Saclay) comes *Index de la Littérature Nucléaire Française*, twice a month; despite its title, abstracts are included. The literature coverage is not confined to CEA publications, and includes reports, theses, patents, conference and journal papers and books.

The *Bulletin Signalétique* (CNRS) covers nuclear science in Section 150: Physique, chimie et technologie nucléaires. *Informationen zur Kernforschung und Kerntechnik*, from ZAED, provides short fast abstracts, in English, of conference papers, dissertations and theses originating from Germany. Items are grouped in accordance with the *INIS* subject categories.

The Russian *Referatjvnyi Zhurnal Fizika* (VINITI) is issued monthly in seven parts, of which B deals with Theoretical Physics and Elementary Particles and V with Nuclear Physics. All the world's literature is covered (about one-quarter of the entries are of Russian origin), the abstracts being all in Russian. A rudimentary check on a small sample of elementary particle papers, by Russian authors, indicates that about 80% of them were included in *NEA* and in *High Energy Physics Index*. On the basis of the nominal publication dates of the three journals, *NFA* was on average 1 month later than *RZh* in listing a given article, while *HEPI* was 1 month early. The order of appearance in practice will depend on relative postal delays. This, and the degree of familiarity with the language, will determine the utility of these national publications; in general, there is not much to be gained by taking them in addition to *NSA*.

In *Physics Abstracts* the subject classifications* have recently been revised. Elementary particle physics, formerly arranged by particle type, is now sub-divided by type of interaction and then by process; this is more in line with current attitudes. Nuclear physics, much as before, is divided into structure and properties, decay and radioactivity, reactions and scattering, properties of specific nuclei, nuclear power. Accelerators and instrumentation are now classed with other (e.g. optical) measurement techniques rather than with the associated physics. Cosmic rays are located with geophysics, astrophysics and astronomy. In *Electrical and Electronics Abstracts* the topic Particle and Radiation Production and Instrumentation has more entries than the corresponding portion of *Physics Abstracts*. This is because the technical aspects of detectors and accelerators (i.e. the engineering and electronics features) are also covered. In borderline cases such as this a paper will often be listed in both these journals.

BIBLIOGRAPHIES

Bibliographies in nuclear physics are numerous and varied in presentation. A few are published as books, such as Kuchowicz's *The bibliography of the neutrino, 1929–1965* (Gordon and Breach). Some very useful lists but without annotation may be found at the end

*ICSU. Abstracting Board. *World physics classification*. London, INSPEC, 1972.

of review articles in such publications as *Annual Review of Nuclear Science*. Many bibliographies are issued in report form and these are best traced under the heading 'Bibliographies' in the subject index to *Nuclear Science Abstracts*; material such as the US Atomic Energy Commissions TID series will be included.

The IAEA issues two or three times a year *List of bibliographies on nuclear energy*, which includes compilations published and in progress. Section 8 contains high-energy physics and Section 9 nuclear physics. The IAEA has also issued a series of bibliographies on nuclear energy, although only a few fall within the scope of this chapter. The bibliographies are compiled by IAEA information staff and are issued in a common format with extensive coverage and an English abstract in almost every case.

The ZAED also issues bibliographies in the series *Bibliographien zur Kernforschung und Kerntechnik*; originally in the series AED-C and now as ZAED-Bibls. Section 8 covers cross-sections and Section 6 nuclear physics; the lists are issued at irregular intervals. The material is taken from *Nuclear Science Abstracts*, and may be compiled by another documentation centre with publishing, distribution and technical editing being undertaken by ZAED.

The National Bureau of Standards has issued several nuclear bibliographies in the Special and Miscellaneous Publication Series which may be traced in the list of publications issued as part of the NBS annual report.

DATA COMPILATIONS

In a subject in which vast amounts of data may result from even one experiment data compilations are important.

An early publication resulted from the Nuclear Data Project of the US National Academy of Sciences–National Research Council which produced *New Nuclear Data* as issue 24B of *Nuclear Science Abstracts from* 1952 to 1956. In 1957 this was issued separately and then included data on elastic and inelastic scattering of charged particles which was back-dated to 1950. This publication continued to 1965, when Academic Press reissued the loose-leaf sheets in bound form. In the same year they commenced the monthly publication *Nuclear Data B* (now *Nuclear Data Sheets*) to gradually replace the original *Nuclear Data Sheets*. *Nuclear Data Tables* (previously *Nuclear Data A*) is a collection of other compilations and evaluations of experimental and theoretical results in nuclear physics and charged particle cross-sections are included in this irregular publication. Academic Press also publish a similar journal called *Atomic Data*; this is edited by Dr. Katherine Way, who has been involved in nuclear data compilation since the original NAS–NRC project.

Another early series was begun in 1955 by Brookhaven National Laboratory[1] as *BNL–325* ('the Barn Book')[1]. This was a second edition of AECU–2040 and included declassified material from the security classified report BNL–250. The aim was to give best values of neutron cross-sections with estimated errors rather than a complete list of all measured values. A second edition was issued in July 1958 with a three-volume supplement in five parts being produced between 1958 and August 1966.

A more restricted compilation is the six reviews of energy levels of light nuclei up to $A=20$[2–7] which have been published in the journal *Nuclear Physics* over the 13 years from 1959 to 1972. The last review updates material in the first review, and Dr. Ajzenberg-Selove states that she expects to begin a new generation of review papers in 1973. International collaboration has produced the Computer Index of Neutron Data (*CINDA*)[8] published from 1973 by the IAEA for the USAEC, the European Nuclear Energy Agency Neutron Data Compilation Centre (ENEA NDCC), who produced earlier annual editions, and the USSR Nuclear Data Centre. The index contains bibliographical references to measurements, calculations and evaluations of neutron cross-sections and other microscopic neutron data. From 1973 IAEA is also taking over responsibility for the international request list for neutron data last issued by ENEA NDCC* as *RENDA 72*; future issues will be called *WRENDA 73*, etc. ENEA NDCC lists in its *CCDN Newsletter* the extensive library of data received from European research centres.

Landolt-Börnstein[9] is justly famous for its careful and complete compilation of data in science and technology. Nuclear physics comprises Group 1 of the new series and there is a gap of 6 years between the publication of the first volume in 1961, on energy levels, and the second volume in 1967, on nuclear radii. The first paper, by Ajzenberg-Selove and Lauritsen, is based on the original article in *Nuclear Physics* and should thus be read in conjunction with the last item from that series. A greater emphasis has been placed on elementary particles, where the aim is to supplement the tables issued by the major laboratories with a critical selection of the data. The first of the two major compilations on elementary particle physics is

---

*The European Nuclear Energy Agency (ENEA) is sponsored by OECD and is now named the Nuclear Energy Agency (NEA) following the inclusion of Japan. The section of NEA called Neutron Data Compilation Centre (NDCC) serviced the international Requests for Neutron Data (RENDA 72) but has now handed over responsibility to the IAEA, who will issue this as Worldwide file of Requests for Neutron Data (WRENDA 73).

issued annually from the Particle Data Group at Berkeley, and by CERN. This Review of Particle Properties covers leptons, mesons and baryons and is issued both as a report and as a review article which alternates between *Reviews of Modern Physics* (1971, 1973) and *Physics Letters* (1970, 1972). The tables present the data as weighted averages with some critical judgement in the form of comments and mini-reviews.

CERN also issue the other compilation series on elementary particles which started from the presentation of data on two-body reactions at Stony Brook Conference, 1966. This was enlarged and became part of the European project of data compilation, High Energy Reaction Analysis (HERA), from which the reports take their numbering code CERN-HERA.[10] The material is regularly updated and is critically evaluated. Data in the field of photonuclear physics is available from and published by the Photonuclear Data Center of the National Bureau of Standards.[11] A biennially updated index to data, with annotations, is produced from the extensive data files, the last cumulation for 1973 covering material from 1955 to 1972. The Center will also deal with individual requests for data and for computer-generated bibliographies. Isotope data information is available from commercial sources such as the Radiochemical Centre, Amersham, which produced the *Radiochemical manual*[12] in 1966 and is supplemented by their annual catalogues of radioactive sources and chemicals.

The valuable series of publications by the Chemical Rubber Co. includes *Handbook of the radioactive nuclides*.[13] Although not as commercially oriented, it is similar to the *Radiochemical manual* in having a part as a narrative presentation and explanation and part in graphical and tabular form.

A chart of nuclides has also been produced by the German Ministry responsible for scientific research.[14]

A continuing series has been the *Table of isotopes*,[15] first issued in 1940 under the editorship of J. J. Livingood and G. T. Seaborg and now in its sixth edition under C. M. Lederer and others.

A special effect in nuclear science is covered in the *Mössbauer effect data index*,[16] which was initially distributed on an informal basis by the North American Aviation Science Center in three volumes covering 1958–64. Later editions to 1971 have been published commercially.

CONFERENCES AND SUMMER SCHOOLS

Meetings of this type are dual-purpose; for the participants they provide opportunities for face-to-face discussions in addition to the formal sessions, and for the scientific community at large they constitute, via the published proceedings, another communication medium. Invited papers are usually major reviews, while the other papers frequently report preliminary results or may be more speculative than journal papers.

Even if a completed piece of work is presented at a conference, this is not regarded as final publication and a journal paper, or a report, should always appear. It follows that conference proceedings have only a short useful life, and in recent years there has been a welcome realisation that rapid publication is vital. The hard-cover, typeset volume, often a year or more late, is giving way to proceedings in report format, offset-lithographed direct from authors' typescripts, and issued within months (and cheaper into the bargain). Coupled with this is a trend for proceedings to be published by the institution that organised the event, rather than by a commercial publishing house. This is, on balance, advantageous for rapid publication, but can give rise to bibliographical problems later. Conference proceedings are treated like reports by *NSA*; conferences are allocated numbers such as CONF 720419 (which means the 19th conference (in arbitrary sequence) held in April 1972). Individual papers (e.g. those issued as preprints) are indexed under this CONF number, as are the proceedings if and when published. Proceedings are listed by other abstract journals as well.

There is one published specialist listing of forthcoming conferences in this field—*Meetings on Atomic Energy* (IAEA). This appears quarterly, and looks ahead for a year or more in conferences and exhibitions, and for a lesser period in training courses.

There is no consistent rule, even within a series, as to what should be included in the published proceedings. Invited papers are invariably included. Contributed papers that were not actually read are rarely printed in full; title and authors are given and occasionally an abstract. What is done with contributed papers that were selected for presentation depends on the format of the conference. If the rapporteur system was used, whereby a talk is given summarising all the contributed papers, even those that were presented may not be reproduced. It is also not uncommon for a paper to be omitted from the proceedings if it is to be published in a journal in substantially the same form.

In elementary particle physics, the leading conferences are those in the 'Rochester' series. Initially held annually at Rochester, NY, USA, these conferences are now held every 2 years under the auspices of IUPAP. The conference takes place in turn in Europe, the USSR and the USA. Invited papers at these conferences are major authoritative reviews, and there is always fierce competition to attend. The

'Aix-en-Provence' series is also based on a 2-year cycle, alternating with the 'Rochester' series. It was originally intended for younger European particle physicists unable to attend Rochester conferences. Although always held in Europe, the attendance now has a non-European component and the age criterion is seldom observed. The 'Aix' series is not formally sponsored. Other series sponsored by IUPAP are concerned with Electron–Photon Interactions and instrumentation for High Energy Physics. The Particles and Fields and the Nuclear Physics Commissions of IUPAP also support a series entitled High Energy Physics and Nuclear Structure; this does not encompass the whole of both these separate subjects but the area common to them. The IUPAP Nuclear Structure Commission also sponsors conferences within its field. The IAEA organise frequent international conferences in various parts of the world; a number of these form continuing series on the same topic, examples being those on Fission and on Nuclear Data for Reactors. Conferences are also frequently organised by the IAEA's International Centre for Theoretical Physics in Trieste.

There are many conference series that are international in the sense of drawing both speakers and participants from many countries, but which take place in the same institution or country. Topics which are catered for in this way include Mössbauer Spectrometry, Cosmic Rays, Nuclear Quadrupole Resonance Spectroscopy, Liquid Scintillation Counting, Radiation Protection, Particle Accelerators. The last-named subject also has two IUPAP sponsored series; they are concerned with high-energy machines and with cyclotrons, and have been held in a variety of particle physics and nuclear physics laboratories.

The United Nations Conferences on the Peaceful Uses of Atomic Energy are highly formal gatherings with national delegations rather than individual participation. It has become (together with the associated exhibition) a shop-window where national prestige is a major consideration. The second conference in 1958 was noteworthy for major survey papers on nuclear fusion research—a previously secret topic. The third conference in 1964 and the fourth in 1971 concentrated largely on nuclear power (particularly the economic and social aspects), the back-up research and development, and the uses of nuclear techniques in industry and agriculture.

The term 'summer school' embraces a wide range of activities, not necessarily confined to the summer season. The distinguishing feature of them all is that they are basically pedagogic; new material may be presented, but only incidentally. The lectures at the more advanced schools are published and often constitute the most up to date review of a topic. A number of centres have become recognised as possessing the desirable characteristics for a summer school venue. Among them are Les Houches (France), Erice (Italy), Herceg Novi (Yugoslavia) and Brandeis (USA). The programmes of these schools are not exclusively devoted to nuclear matters. A well-established series is the Enrico Fermi School, held at Varenna under the auspices of the Societ à Italiana di Fisica. There are about three courses per year, of which one or two are on nuclear subjects.

PREPRINTS AND REPORTS

Preprints

The term 'preprint' was coined to denote the multiply copied typescript of a paper, destined for private circulation in advance of formal publication. This practice grew up in an effort to by-pass the communication time-lags that resulted from formal publication, complete with referring and typesetting delays. The phenomenon has been noted in other rapidly advancing subjects. Inevitably the original concept has been outgrown; preprints are distributed widely by means of 'standard' lists of institutions and individuals, and it by no means follows that material in a preprint is intended for publication. It is often used, particularly by theorists, to publicise undeveloped ideas or preliminary results; since material of this type is liable to subsequent retraction, the authors are often reluctant to issue their preprints within official institutional documentation systems. This makes bibliographical discipline difficult to impose and often leads to publications being un-numbered and unknown to the library of the parent organisation. There is also the problem of the closed circle—the information being available, in the first instance at least, only to those persons and institutions to whom the author has seen fit to send it. Some measure of order has been introduced by services such as *Preprints in Particles and Fields* (see Current Awareness Journals, above) with its Antipreprint lists giving the open literature references of those preprints (about 80%) that are eventually published. The coverage of preprints by abstract journals is very variable; in elementary particle physics there is very good coverage by PPF and HEPI, and understandably comparatively little by *NSA* and hardly any by *PA*.

Reports

Reports have always been an important part of the literature of nuclear subjects; they are very suitable for recording information not suited to conventional publication for reasons of confidentiality or size (e.g. large quantities of tabular data). The availability of reports depends on the contents and on the originating laboratory. Those which are freely available are frequently distributed to other institutions on an exchange basis, and are usually available on request to bona fide individual requestors, or, on a commercial basis, from the National Technical Information Service, Springfield, Va.

No useful purpose would be served by giving extensive lists of the identification codes used by the institutions that issue reports. These usually consist of a string of letters identifying the institution, followed by a serial number that sometimes embodies the year of issue. A valuable aid is the *Dictionary of report codes*.[41] The indexing of reports by *NSA* is very good. Each half-monthly part has a report number index which, in addition to giving the abstract number, also lists the availability, number of pages and price. Further information is given in the preamble to the report number index. The cumulations of this index include, in addition, a list of all the report number series used in *NSA* together with the name and address of the originating institution; there are about 6000 entries. The corporate author index of *NSA* gives the number codes used by each organisation. Most reports are now available on microfiche; where the option exists microfiche is substantially cheaper and quicker to obtain than hard copy.

Nuclear science reports are also listed in the general compilations such as *R and D Abstracts* (DTI Technology Reports Centre, St. Mary Cray) and the *US Government Reports Announcements and Index* (National Technical Information Service), which was formerly *US Government Research and Development Reports*.

BOOKS

Although intended for use in conjunction with other material the books of the *Open University Science Foundation Course units 31, 32*[17] provides a useful introduction with questions and answers but no bibliographical references. A clearly written popularisation of a different type is Gamow's *The atom and its nucleus*,[18] but again without bibliographical references. Somewhat more advanced introductory tests are Kaplan's *Nuclear physics*[19] and the new edition of Burcham's *Nuclear physics: an introduction*.[20] An early though comprehensive coverage of work and literature to 1953 is *Experimental nuclear physics*, by Segré,[21] and another classic text is Born's *Atomic physics*, with later editions revised by R. J. Blin-Stoyle.[22] *Introduction to atomic and nuclear physics*, by White,[23] is a well-produced book with descriptions of the actual experiments; this same emphasis is found in a clearly written book by Yarwood, *Atomic and nuclear physics*.[24] The experimental approach is also favoured by Marmier and Sheldon in their *Physics of nuclei and particles*[25] in two volumes. The first covers properties of nuclei and the second interactions of nucleons; fundamental particles are to be included in a projected third volume. More specialist texts in the fields of nuclear reactions and structure are Jackson's *Nuclear reactions*,[26] compiled from her lectures to postgraduate students; a reprint volume of classic papers, *Nuclear reactions*,[27] edited by McCarthy; and the cautious but scholarly *Nuclear reactions and nuclear structure*,[28] by Hodgson.

The field of elementary particles has produced many expensive books for specialists. This is not an easy subject to popularise but two books which can be recommended are Matthews's *The nuclear apple*[29] and Wick's *Elementary particles*.[30] Some knowledge of quantum mechanics is assumed in three works of a similar level and standard: Hughes's *Elementary particles*;[31] Perkins's *Introduction to high energy physics*;[32] and the second edition of Williams's *An introduction to elementary particles*.[33] The subject is given a more extensive treatment in Paul's *Nuclear and particle physics*,[34] although the most comprehensive work is *High energy physics*,[35] edited by Burhop. Originally intended as a three-volume compilation, this has grown to five volumes with the last volume including a revision of a paper in volume 2.

The final book covering the basic physics, edited by C. Weiner,[36] is a somewhat unusual item made from the transcript of two small conferences when leading nuclear physicists brought together by the American Institute of Physics offered recollections and interpretations of historical events in which they, and others now dead, had participated.

On the instrumentation side it is somewhat hazardous to recommend books on account of the frequent rapid advances in particle detection techniques. *Bubble and spark chambers*, edited by R. P. Shutt,[37] and *Semiconductor counters for nuclear radiations*, by Dearnaley and Northrop,[38] cover techniques which are still in use albeit with improvements in detail. Livingston and Blewett's *Particle accelerators*[39] and Livingood's *Principles of cyclic particle accelerators* are still sound despite being over 10 years old.

DIRECTORIES, DICTIONARIES AND ENCYCLOPAEDIAS

The *World nuclear directory* (Harrap) is now in its 4th edition. Under each country details are given of the nuclear activities of government departments and agencies, universities, professional and learned societies, and industry. Entries are liberally supplied with names of directors and departmental heads. Lists of high-energy physics institutes are issued by SLAC as a by-product of their *PPF* service and also by CERN. The IAEA publish a directory of nuclear reactors. A recent venture is *Nuclear Research Report* (Nuscience Publications) which is a 'compilation of on-going research projects'. Issued yearly, it lists nuclear and high-energy research by country and institution.

There are a number of multi-language dictionaries and glossaries covering nuclear science. Two in the Elsevier series are the *Dictionary of nuclear science and technology* and the *Dictionary of electronics, nucleonics and telecommunications*. The languages covered are English/American, French, Spanish, Italian, Dutch and German. There is a French and English *Electronics and nuclear physics dictionary* (Dunod), a Russian–English *Nuclear dictionary* (Fitzmatgiz, 1960), and an English to Russian *Dictionary of charged particle accelerators* (Publishing House Sovietskaia Entsiklopedia, 1965).

Glasstone's *Sourcebook on atomic energy* (van Nostrand, 1967), now in its 3rd edition, is a valuable reference tool. It gives basic information covering the whole subject, presenting the underlying science in a way that can be understood by the non-specialist). The monumental *Encyclopedia of physics* (*Handbuch der Physik*. Springer) (see p. 37) gives good coverage to nuclear science. The volumes are grouped by subject matter. Group VI (Röntgen and Corpuscular Rays) contains four volumes of interest to nuclear scientists and there are eight volumes in Group VIII (Nuclear Physics). These 12 volumes contain over 60 articles, of which about 90% are in English, the rest in German. There are both English and German subject indexes. The work is now over 10 years old and shows its age, particularly in the articles on instrumentation.

REFERENCES

1. Hughes, D. J. and Harvey, J. A. (1955). Brookhaven National Laboratory, Upton, N.Y. *Neutron cross sections*. BNL-325 Suppl. 1, Jan 1957; 2nd edn, July 1958; Suppl 1, Jan 1960; Suppl 2 v 1–3, 1964–66
2. Ajzenberg-Selove, F. and Lauritsen, T. (1959). *Nuclear Physics*, 11, 1–340
3. Lauritsen, T. and Ajzenberg-Selove, F. (1966). *Nuclear Physics*, 78, 1–176
4. Ajzenberg-Selove, F. and Lauritsen, T. (1968). *Nuclear Physics*, 114, 1–142
5. Ajzenberg-Selove, F. (1970). *Nuclear Physics*, A152, 1–221
6. Ajzenberg-Selove, F. (1971). *Nuclear Physics*, A166, 1–139
7. Ajzenberg-Selove, F. (1972). *Nuclear Physics*, A190, 1–196
8. CINDA. *Computer Index of Neutron Data*. Vienna, International Atomic Energy Agency, 1967– . Annual

9. Landolt-Börnstein. For detailed description see p. 41.
10. CERN HERA-69-1: European Organisation for Nuclear Research, Geneva. *A compilation of pion–nucleon scattering data*
    CERN-HERA-69-2: *Data compilation of antiproton–proton into anti-hyperon–hyperon*
    CERN-HERA-69-3: *A compilation of total and total elastic cross sections*
    CERN-HERA-70-1: *A collection of pion photoproduction data. I. From threshold to 1·5 GeV*
    CERN-HERA-71-1: *Selective compilation of $\pi p^- \to \pi\pi N$ events from hydrogen bubble chambers*
    CERN-HERA-72-1: *Compilation of cross sections. I. π^- and π^+ induced reactions.* Updated version of CERN-HERA-70-5
    CERN-HERA-72-2: *Compilation of cross sections. I. K^- and K^+ induced reactions.* Updated version of CERN-HERA-70-4
    CERN-HERA-73-1: *Compilation of cross sections. III. p and p induced reactions.* Updated version of CERN-HERA-70-3 and 70-3
11. Fuller, E. G. *et al.* (1973). *Photonuclear reaction data.* US National Bureau of Standards Special Pub. 280
12. Wilson, B. J. (ed.). *Radiochemical Manual.* 2nd edn, Radiochemical Centre, Amersham
13. Wang, Y. (ed.). *Handbook of the radioactive nuclides.* Chemical Rubber Co.
14. Seelmann-Eggebert, W. *et al.* (eds). (1968). *Nuklidkarte.* 3rd edn, Bundes-minster fur Wissenschaftliche Forschung, Bonn
15. Lederer, C. M. *et al.* (eds.) (1967). *Table of isotopes.* 6th edn, Wiley
16. Mössbauer effect data index. 1958–65, ed. Muir A. H. *et al.* Interscience, 1966
    1969, ed. Stevens, J. G. and V. E. Plenum, 1970
    1970, ed. Stevens, J. G. and V. E. Plenum, 1971
    1971, ed. Stevens, J. G. and V. E. Plenum, 1972
17. Open University (1971). Science Foundation Course Unit 31: *The nucleus of the atom.* Science Foundation Course Unit 32: *Elementary particles.* Open University
18. Gamow, G. (1961). *The atom and its nucleus.* Prentice-Hall
19. Kaplan, I. (1963). *Nuclear Physics.* 2nd edn, Addison-Wesley
20. Burcham, W. E. (1973). *Nuclear physics; an introduction.* 2nd edn, Longmans
21. Segré, E. *Experimental nuclear physics.* Vols 1–3, Wiley (1953–1959).
22. Born, M. (1962). *Atomic physics.* 7th edn, rev. by the author in collaboration with R. J. Blin-Stoyle, Blackie
23. White, H. E. (1964). *Introduction to atomic and nuclear physics.* Van Nostrand
24. Yarwood, J. (1973). *Atomic and nuclear physics.* University Tutorial Press
25. Marmier, P. and Sheldon, E. (1969, 1970). *Physics of nuclei and particles.* Vols. 1, 2, Academic Press
26. Jackson, D. F. (1970). *Nuclear reactions.* Methuen
27. McCarthy, I. E. (ed.). (1970). *Nuclear reactions.* Pergamon
28. Hodgson, P. E. (1971). *Nuclear reactions and nuclear structure.* OUP
29. Matthews, P. T. (1971). *The nuclear apple: recent discoveries in fundamental physics.* Chatto and Windus
30. Wick, G. L. (1972). *Elementary particles: frontiers of high energy physics.* Geoffrey Chapman
31. Hughes, I. S. (1972). *Elementary particles.* Penguin

32. Perkins, D. H. (1972). *Introduction to high energy physics.* Addison-Wesley
33. Williams, W. S. C. (1971). *An introduction to elementary particles.* 2nd edn, Academic Press
34. Paul, E. B. (1969). *Nuclear and particle physics.* North-Holland
35. Burhop, E. H. S. (ed.). (1967–1972). *High energy physics.* Vols 1–5, Academic Press
36. Weiner, C. (ed.). (1972). *Exploring the history of nuclear physics: proceedings of the American Institute of Physics–American Academy of Arts and Sciences Conferences on the History of Nuclear Physics, 1967 and 1969.* AIP Conference Proceedings No. 7
37. Shutt, R. P. (ed.). (1967). *Bubble and spark chambers* (3 vols.). Academic Press
38. Dearnaley, G. and Northrop, D. C. (1966). *Semiconductor counters for nuclear radiations.* 2nd edn, Spon
39. Livingston, M. S. and Blewett, J. P. (1962). *Particle accelerators.* McGraw-Hill
40. Livingood, J. J. (1961). *Principles of cyclic particle accelerators.* Van Nostrand
41. Godfrey, L. E. and Redman, H. F. (1973). *Dictionary of report codes.* 2nd edn, Special Libraries Association, 1973

14

Crystallography

A. L. Mackay

Few fields of science are as well organised as is crystallography, although, with interest in the solution of crystal structures as a problem in itself now past its apogee, there are signs of diversification and fragmentation in a hitherto monolithic edifice. It is appearing that the subject matter of crystallography is only coincidentally crystals. The real subject matter is the structure of matter at a level above that of the structure of the individual atom and below that of the reaction system. The techniques of microscopy, the use of waves of light, X-rays, electrons and neutrons, and most lately also protons, to see these structures, of course form part of the subject, as does the technology of the computing which transforms the data and acts as the lens of the generalised microscope.

A great part of present-day crystallography supplies results which are essential to other branches of science, such as chemistry, biology and solid state physics, and consequently the production of compilations of data is especially well-developed, although there is a tendency, facilitated by modern computers, towards the provision by data services of answers to individual questions. In spite of such tendencies the book is holding its own.

The International Union of Crystallography, affiliated to the International Council of Scientific Unions, was founded in 1946, but before that there were co-operative projects for the rational publication of crystallographic data. Adherence to the IUC is through national crystallographic organisations (in the case of the UK, the British National Committee for Crystallography, which is operated by the Royal Society). The IUC itself publishes a series of books and periodicals. In this survey questions of mineralogy, other than the structures of crystals, are largely excluded, although crystallography is fundamentally interlinked with mineralogy as it is with a number of other subjects. Crystallography has outgrown its origins as a section of mineralogy.

HISTORICAL

The present X-ray period, starting from the classic experiment of von Laue, Friedrich and Knipping (1912), is well documented by P. P. Ewald (*59 Years of X-ray diffraction*, IUC, 1962) in a book produced to mark the 50th anniversary. Ewald was an active participant in the events leading to the discovery of X-ray diffraction and, at the time of writing, is still active.

The earlier stages of the studies of the structures of crystals—from Pliny and Theophrastus to von Laue—are covered by J. G. Burke (*Origins of the science of crystals*, University of California Press, 1966), who furnishes many references facilitating further investigation. There is also a growing hagiography covering key figures such as Fedorov, Bravais and Pauling. The last-named was presented with a splendid Festschrift for his 65th birthday (*Structural chemistry and molecular biology*, ed. A. Rich and N. Davidson, Freeman, 1968). There are several histories of crystallography in its pre-atomic guise of mineralogy and of the discoveries of Pasteur and others. The account of the discovery of the structure of DNA by J. D. Watson (*The double helix*, Weidenfeld and Nicolson, 1968) carries history into the present. The IUC has produced two volumes of *Early papers on diffraction of X-rays by crystals* (ed. J. Bijvoet, W. G. Burgers and G. Hägg, 1972) which make the history accessible. Shafranovskii has published a history of crystallography in Russia which contains many references for deeper research *Istorija Kristallografiiv Rossii*, Leningrad, Nauka, 1962).

The studies of the history of the science of crystals outside Europe is still fragmentary, but token mention might be made of a facsimile edition of 'Sekka-zusetsu' illustrations of snow crystals by Doi Toshitsura (1789–1848) with commentary by Kobayashi Teisaku (1968). Joseph Needham, of course, gives an excellent lead in to the history of the early study of crystals in China. Post-X-ray history in Japan was reviewed by Nitta Isamu (*Acta Cryst.*, **A29**, 315–322, 1973).

CRYSTALLOGRAPHIC BOOK LISTS

The IUC maintains a Commission on Crystallographic Teaching and in 1965, under the editorship of Dr. Helen Megaw, a comprehensive *Crystallographic book list* was published. This covered books in English, French, German, Spanish and Russian, classifying them by level and by topic and giving an account almost complete up to 1964. (The only major omission is of books in Japanese, the majority of which are either translations of material which has appeared elsewhere or texts similar to those available in English. The Japanese publish increasingly in English for their more original material.) A supplement to the book list followed in 1966.

The *Crystallographic book list* is comprehensive in including serials and conference proceedings as well as books. Supplements will be produced in the future and M. M. Woolfson published the Second in *J. Appl. Cryst.*, **5**, 148–162 (1972). This included references to reviews, so that the inquirer can consult an informed opinion about any book. To some extent, then, the present review must be a personal exegesis of this list.

PERIODICALS

As in almost all other subjects, the most active part of the literature consists of papers in regular periodicals. The principal journals devoted to crystallography are the following, although many crystallographic papers appear elsewhere:

Acta Crystallographica A & B (IUC)
Journal of Applied Crystallography (IUC)
Zeitschrift fuer Kristallographie (Akademische Verlagsgesellschaft, Frankfurt-am-Main)
Kristallografiya (Acad. Sci. USSR) (in English translation as: *Soviet Physics: Crystallography*, American Institute of Physics)
Journal of Crystal and Molecular Structure (Plenum)
Zhurnal Strukturnoi Khimii (Acad. Sci. USSR) (in English translation as: *Journal of Structural Chemistry*, Consultants Bureau)
X-sen [*X-rays*] (Japanese)
Bull. de la Societé française de Mineralogie et Cristallographie (Masson)
Journal of Crystal Growth (North-Holland)
Crystal Research and Technology (Akademie, Berlin)
Bulletin Signalétique: Cristallographie (CNRS)
Molecular Crystals and Liquid Crystals (Gordon and Breach)

SERVICES

While the collections of crystallographic data in book form are well organised, the increasing tempo of the production of structures has meant that they fall behind in their reports on the primary literature (in journals). A number of services aiming for more rapid analysis are active.

The principal of these is the Crystal Data Centre, directed by Dr. Olga Kennard at the University Chemical Laboratory, Cambridge. This is supported by OSTI and was founded on the initiative of J. D. Bernal. Here all papers dealing with organic and organometallic structures are recorded in machine-readable form. The internal consistency of structures is checked and bibliographies of the papers are now published regularly, as are digests of bonding information (*Molecular structures and dimensions*, Vols. 1, 2, 3, 4 (bibliography 1935–71); Vol. A1 (dimensions), ed. O. Kennard and D. G. Watson, Crystal Data Centre Cambridge and IUC). A number of searching procedures are available on application.

A similar centre covering the inorganic structures is in operation at McMaster University (Hamilton, Ont.) under the direction of Prof. I. D. Brown.

The metallurgical literature centres on Dr. W. B. Pearson (National Research Council, Ottawa), who has published the monumental data books: *A handbook of lattice spacings and structures of metals and alloys* (Pergamon, 2 vols. 1958 and 1967).

As part of the larger literature searching and abstracting schemes, the CNRS (Paris) will retrieve crystallographic papers and their Bulletin Signalétique (Section 161: Cristallographie, formerly Structure de la Matière) may be reckoned as the main abstracting journal in the field.

The Institute of Scientific Information (Philadelphia, USA) provides commercially services based on their machine file of the *Science Citation Index*.

A number of minor agencies and societies provide abstracting services. We should mention, for example, *Mineralogical Abstracts* (Mineralogical Society, London).

Protein structures are collected by the Crystal Data centre at Cambridge but this is also done by the National Biomedical Research Foundation (Georgetown University Medical Center, Washington, D.C.) who publish the *Atlas of protein sequence and structure* (now at Vol. 5, 1972), edited by Margaret Dayhoff, which is an indispensible compendium (and textbook).

SERIALS

There are numerous serials which deal with crystallography, widely interpreted, notably:

Progress in Stereochemistry (ed. B. J. Aylett and M. M. Harris) (Butterworths Vol. 4, 1969)

Perspectives in Structural Chemistry (ed. J. D. Dunitz and J. A. Ibers) (Wiley, Vol. 3, 1971)

Advances in X-ray Analysis (ed. W. M. Mueller) (Plenum Press, Vol. 12, 1969)

Advances in Structure Research by Diffraction Methods (ed. R. Brill and R. Mason, Interscience, 1964–   )

Progress in Biophysics and Molecular Biology (Pergamon, Vol. 21, 1971)

Transactions of the American Crystallographic Association (ACA)

GEOMETRY, SYMMETRY AND MATHEMATICS

A sound knowledge of geometry is the basis of any study of structure. Euclid's *Elements*, indeed, is still essential but is best approached through H. S. M. Coxeter's masterly account *Introduction to Geometry* (1961 and later edition, Wiley). In the earliest stages one can hardly do better, for an appropriate mathematical grounding, than to work through the volumes of the *School mathematics project* (CUP, 1965–   ), which are replete with concepts relevant to structure and symmetry.

As always, one cannot have too much mathematics and we have found the *Mathematical handbook* of G. A. and T. M. Korn (2nd edn, McGraw-Hill, 1968) to supply most of what is needed. *Mathematical tables*, ed. by J. S. Kasper and K. Lonsdale, 1959 (Vol. 2 of *International tables for X-ray crystallography*, Kynoch Press, Birmingham) contain a summary of most of the mathematics commonly encountered in crystallographic work.

As almost every book on crystallography has some account of symmetry, and most are necessarily explanations of the *International Tables*, there is little need to mention any particular book, although the texts by M. J. Buerger (*X-ray crystallography*, Wiley 1942; *Elementary crystallography*, Wiley, 1956; *Vector space*, Wiley, 1959; *Crystal structure analysis*, Wiley, 1960; *Contemporary crystallography*, McGraw-Hill, 1970) are the most authoritative accounts. For the formal geometrical crystallography *An introduction to crystallography*, by F. C. Phillips (Longmans, 1963) remains unsurpassed.

Crystallography is an area of culture shamefully neglected by most non-scientific commentators, although reference to the Bible, Pythagoras, Plato, Euclid, Leonardo da Vinci, Dürer, is sufficient to show that ideas of mathematical symmetry are fundamental to our civilisation. However, there are at least two books which examine the importance of mathematical symmetry in general culture: these are Hermann Weyl's *Symmetry* (Princeton UP, 1951) and *Symmetry in science and art* (Russian), by A. V. Shubnikov and V. A. Koptsik (Nauka, Moscow, 1972). The prints by M. C. Escher have achieved increasing popularity, particularly in a definitive collection *The World of M. C. Escher* (Abrams, 1971), and in an edition by Caroline MacGillavary: *Symmetry aspects of M. C. Escher's periodic drawings* (Int. Union Cryst., 1965), which is designed for the entertainment and instruction of crystallographers.

This is perhaps also the place to recommend most highly D'Arcy W. Thompson's classic *Growth and form* (CUP, 1942); (although we should repeat Coxeter's warning that Haeckel's beautiful drawings of symmetrical Radiolaria, so frequently reproduced, were most probably invented by him rather than drawn from life).

The orthodox mathematical literature is disappointing to the crystallographer. The topics in which he is interested were 'worked out in the nineteenth century' and are forgotten and neglected by the present-day mathematician, so that the crystallographer must exhume what he wants or redevelop for himself. Mathematically, the great struggle at present is for a theory which will describe chance of state adequately. There are already signs (*Science*, **181**, 147, 13 July 1973) that a universal theory of critical phenomena will come from extensions of group theory and geometry to four dimensions. Tables of extensions of space-group theory have been published by V. A. Koptsik (*Shubnikov groups*, Moscow University, 1966) representing the work of a group of Russian crystallographers, but as the material will be digested into the new edition of the *International tables* (in preparation) it is unnecessary to consult it. The present edition of the *International tables for X-ray crystallography* (3 vols. Kynoch Press, Birmingham, 1952, 1959, 1962) is the crystallographer's Bible, supplying him with essential data on notation, symmetry, constants and almost-all other basic items.

COMPUTING

The field of computing is still much of a jungle with special local rules for the implementation of a program on any particular machine. Apart from characteristics of the operating systems necessitating special job descriptions, individual machines may use compilers with local 'dialects' of the major programming languages—Fortran, Algol, Basic, etc. There is some effort now to use standard versions such as ASA Fortran, which may eschew certain short-cuts but which are universally acceptable. The prime texts must, therefore, be the local computer manuals.

The major suite of computer programmes is called X-ray 70 (and is in Fortran). It is well documented by its chief architect, J. M. Stewart, and is available at many major computer centres. Most centres keep a tape library of the programmes they have collected. 'A world list of crystallographic computer programs appeared in *J. Appl. Cryst.*, **6**, Pt. 4, 309–346 (1973), and supersedes previous lists.

There are really only a few books in wide use—for example, *Computing methods in crystallography*, edited by J. S. Rollett (Pergamon, 1965), and *Crystallographic computing*, edited by F. R. Ahmed (Munksgaard, 1970)—although more must be expected with the intensive automation of all departments of the subject.

Very many books of tables have been made obsolete by the rise of computing power in the shape of individual hand calculators, time-sharing links and large-capacity batch processing. Visual display devices of all kinds are proliferating and all these have their manuals.

DIFFRACTION THEORY

The main features of diffraction theory were worked out remarkably fully by the pioneers of X-ray crystallography, such as Ewald, von Laue and C. G. Darwin, but their works are little used, most readers preferring more modern representation. With the advent of the laser, X-ray interferometry, radio astronomy and electron microscopy there has been a renaissance in physical optics and all wavelengths are seen as showing similar phenomena in their interaction with periodic structures.

The works of H. S. Lipson and his school are widely read—for example, *Optical physics*, by S. C. and H. Lipson (CUP, 1969), and *Optical transforms*, by C. A. Taylor and H. Lipson (Bell, 1964). *The optical principles of the diffraction of X-rays*, by R. W. James (his *The crystalline state*, Vol. 2, Bell, 1963) is the most used text on diffraction theory as regards X-rays.

For electron diffraction there is a mass of material of recent origin, much emanating from the very strong Japanese school of electron microscopy, appearing mainly in current periodicals.

Molecular crystals, their transforms and diffuse scattering, by J. L. and M. Amorós (Wiley, 1968), is a good account of thermal vibrations in crystals and its consequences for diffraction from them.

B. K. Vainshtein in *Diffraction of X-rays by chain molecules*

(Russian, Acad. Sci. USSR, 1963; English, Elsevier, 1966) deals with the special problems of the helices encountered in DNA and protein structures.

STRUCTURE ANALYSIS AND GENERAL CRYSTALLOGRAPHY

This is really the central area of crystallography and the present period is dominated by the direct methods developed by Sayre, Zachariasen, Karle and Hauptmann, Woolfson and others which are now implemented in computer programs, which, rather than books, represent the state of the art. Books, however, are still important—for example, M. M. Woolfson's *Direct methods in crystallography* (OUP, 1961) and A. I. Kitaigorodskii's *Theory of crystal structure analysis* (Russian, 1957; English, Consultants Bureau, 1961), which represented an early stage, being written before computers were so fast.

Of the earlier textbooks a number have lasted well, among them C. W. Bunn's *Chemical crystallography* (2nd edn, OUP, 1961) and also W. L. Bragg's *The crystalline state*, vol. 2, *General survey* (Bell, 1933), a remarkable textbook filled with matter worth reading 40 years later. Volume 3 of this series, *The Determination of crystal structures* (Bell, 1966), by H. S. Lipson and W. Cochran, although largely pre-computer, is still a basic account. Of the later books, a really excellent introductory text is J. P. Glusker and K. N. Trueblood's *Structure determination by single X-ray diffraction* (OUP, 1972). Somewhat fuller books are G. H. Stout and L. H. Jensen's *X-ray structure determination* (Macmillan, 1968) and M. M. Woolfson's *An introduction to X-ray crystallography* (CUP, 1970). *Advanced methods of crystallography*, edited by G. N. Ramachandran (Academic Press, 1964), and G. N. Ramachandran and R. Srinivasan's *Fourier methods in crystallography* (Wiley/Interscience, 1970) explain the imaging techniques which are the main alternative to direct methods.

Books which are more concerned with the actual techniques for obtaining information are L. V. Azaroff's *Elements of X-ray crystallography* (McGraw-Hill, 1968) and J. W. Jeffrey's *Methods in X-ray crystallography* (Academic Press, 1971). On the particular problems of the four-circle automatic diffractometer U. W. Arndt and B. T. M. Willis's *Single crystal diffractometry* (CUP, 1966) is authoritative and also covers the collection of neutron diffraction data. Neutron crystallography is presented by two pioneers in the field: G. E. Bacon's *Neutron diffraction* (OUP, 1962) and Y. U. Izyumov and R. P. Ozerov's *Magnetic neutron diffraction* (Plenum, 1970).

CRYSTAL CHEMISTRY

The discussion of the results of structure analysis clearly falls into the field of chemistry but there are a number of books written from the crystallographer's point of view. Most notably these are Linus Pauling's classic *The nature of the chemical bond* (3rd edn, Cornell University Press, 1960) and the very large volume, A. F. Wells's compilation *Structural inorganic chemistry* (3rd edn, OUP, 1962). There are a number of more specialised books—for example, R. G. Burns's *Mineralogical applications of crystal field theory* (CUP, 1970). R. C. Evans's *An introduction to crystal chemistry* (CUP, 1964) was a pioneer systematisation of the structures of crystals representing the views of the pre-war Cambridge school. It was much translated and in its second edition is still recommended as a beginning.

CRYSTAL PHYSICS

Crystal physics was a field which languished for many years, becoming almost confined to the physics of quartz, but, with the establishment of the theory of dislocations and defects, largely using electron microscopic evidence, and the rise of solid state physics, is now very active.

The physics of anisotropic media is well covered by J. F. Nye's, *Physical properties of crystals* (OUP, 1957) and by S. Bhagavantam and T. Venkatarayudu's *Theory of groups and its applications to physical problems* (Academic Press, 1969). The pioneer text in this field was W. A. Wooster's *A textbook of crystal physics* (CUP, 1938), but the same author has superseded it with a study of the effects of thermal vibrations and certain defects on X-ray scattering with *Diffuse X-ray reflections from crystals* (OUP, 1962), which complements Amorós' book mentioned earlier. Wooster also has the latest book as well as the earliest with *Tensors and group theory for the physical properties of crystals* (OUP, 1973).

The most general textbook is *Crystal physics*, by G. S. Zhdanov (Oliver and Boyd, 1965; original Russian, 1961). Crystal optics is a long-established part of crystal physics. Substantial texts are P. Gay's *Introduction to crystal optics* (Longmans, 1967) and N. H. Hartshorne and A. Stuart's *Practical optical crystallography* (Arnold, 1964) and *Crystals and the polarising microscope* (Arnold, 1970).

IMPERFECTIONS, MORPHOLOGY AND GROWTH

The literature in this field is large and diffuse, spreading into metallurgy, solid state physics, chemistry and elsewhere. We should step out of our way to recommend J. E. Gordon's *The new science of strong materials* (Penguin, 1968) as a demonstration that a textbook can be attractive, which is the main purpose of an introduction. Beyond that there is the classic by P. B. Hirsch, A. Howie, R. B. Nicolson, D. W. Pashley and M. J. Whelan, *Electron microscopy of thin crystals* (Butterworths, 1965), where the pioneers of dislocation produced the first substantial book on the new area. We have also A. H. Cottrell's *Theory of dislocations* (Wiley, 1964), J. Friedel's *Dislocations* (Addison-Wesley, 1964) and F. R. N. Nabarros' *Theory of crystal dislocations* (OUP, 1967)—good expositions of this area, where appreciation of the essentially three-dimensional nature of the phenomena is important.

ELECTRON MICROSCOPY AND DIFFRACTION

The electron diffraction analysis of the structures of crystals was pioneered chiefly by the Moscow schools whose books, Z. G. Pinsker's *Electron diffraction* (Butterworths, 1953; original Russian, 1949) and B. K. Vainshtein's *Structure analysis by electron diffraction* (Pergamon, 1964; original Russian, 1956) remain authoritative accounts. The simple electron diffraction camera used by Finch at Imperial College, London, and by Vainshtein and Pinsker has been replaced by the electron microscope, in which the Japanese and the Australian school of J. M. Cowley led the way.

The Japanese school did not seem to publish much as books but *Crystals and waves. Collected papers of S. Miyake* (ed. G. Honjo, Tokyo, 1972) gives a historical sample. The field has advanced so rapidly that conference proceedings must stand instead. The latest is perhaps *Microscopie électronique* (ed. P. Favard) (Resumé des Communications presentées au de Congrès International Grenoble, 1970), which consists of three very large volumes of author's abstracts, excellently illustrated.

BIOLOGICAL STRUCTURES

The application of X-ray and electron microscope methods to biological structures is proceeding at a tremendous rate as the sheer weight of the principal periodical *The Journal of Molecular Biology*

(Academic Press) attests. As yet there are few books by individual authors. Serials such as the *Cold Harbor Symposium on Quantitative Biology* (Cold Spring Harbor Laboratory of Quantitative Biology, No. 1, 1933 to N. 37 1973) and the Nobel Symposium 11, *Symmetry and Function of Biological Systems at the Macromolecular Level* (ed. A. Engström and B. Strandberg, Wiley, 1969) are authoritative but quickly overtaken. *Conformation of Biopolymers*, edited by G. N. Ramachandran (2 vols., Academic Press, 1967) is a similar conference report.

Textbooks age less rapidly, and K. C. Holmes and D. M. Blow's *The use of X-ray diffraction in the study of protein and nucleic Acid structure* (Interscience, 1966), *Diffraction of X-rays by proteins, nucleic acids and viruses,* by H. R. Wilson (Arnold, 1966), and especially the well-illustrated *The structure and action of proteins,* by R. E. Dickerson and I. Geis (Harper and Row, 1969) are strongly recommended.

DATA BOOKS

It will be sufficient to list the principal data compilations not already mentioned since their titles speak for themselves: *World directory of crystallographers,* edited by G. Boom (4th edn, IUC, 1971); *Index of crystallographic supplies* (3rd edn, IUC, 1972); and *Technik-Wörterbuch Kristallographie. Englisch-Deutsch-Französisch-Russisch.* (Backhaus, Berlin, 1972) (there is also an extensive word list in English, French, German, Spanish and Russian in the *International tables,* vol. 2).

For data on unit cell dimensions there is: *Crystal data determinative tables* (3rd edn, 2 vols., ed. J. D. H. Donnay and Helen M. Ondik).

Crystal structures, by R. W. G. Wyckoff, U.S. Dept of Commerce, NBS (5 vols., Interscience) is a private compilation, now in its second edition, which, in book form (the first edition was in looseleaf and was difficult to use) gives convenient descriptions of basic crystal structures. It is, of course, far behind in time.

The *Structure Reports* series of the IUC are also about 10 years behind in the complete reports on all aspects of structural relevance which it aimed to provide. The last volume is Vol. 29 for 1964.

The powder diffraction data file published by the American Society for Testing Materials is essential for industrial work dealing with powder methods of diffraction. The data are available in book form, as cards (notched or plain) and, more recently, as a magnetic tape file. It is very expensive but may be indispensable.

There are a number of handbooks dealing with minerals which have spread somewhat and are most valuable: *The system of mineralogy*, by Dana (Wiley), more an institution than a book; *Mineralogische tabellen*, by H. Strunz (4th edn, Akademische Berlagsgesellschaft, Leipzig, 1966); *Microscopic characters of artificial inorganic solid substances* (A. N. Winchell and H. Winchell, Academic Press, 1964); and *The optical properties of organic compounds* (A. N. Winchell, 2nd edn, Academic Press, 1954).

15

Instrumentation

B. M. Rimmer

Instrumentation equipment is rarely developed and used exclusively for a particular application. It would not be economical to do so. The basic parameters—time, distance, temperature, flow, frequency, etc. —are common to many disciplines. Physicists will, therefore, often be using or adapting equipment devised by engineers or chemists and many will profit from the advanced techniques developed for space research.

In his book on transducers published in 1959, Lion stated that 'few fields of technology rival instrumentation in the almost endless variety of methods and choices available'. This is just as true today but now very much more information is being published. Equipment is closely allied to measuring techniques and the physicist with instrumentation in mind may need to scan material well outside his own subject field. The titles selected for mention here cannot be considered to comprise a comprehensive list. They should serve to give the reader a quick appraisal of the subject and to direct his search into the most suitable area for his particular need.

BOOKS

Instrumentation systems—general

The physicist should be cautious about ordering literature on the basis of the words 'instrument' or 'instrumentation' appearing in the title, as the terms are equally applicable to industrial process control and the equipment is quite different and has little direct relevance to physics.

The following books are suggested:

Electronic instrumentation and measurement techniques (W. D. Cooper. Prentice-Hall, 1970)

Electronic measurements and instrumentation (B. M. Oliver and J. M. Cage. McGraw-Hill, 1971)

These two books could almost be considered companion volumes. Their titles have been carefully chosen to indicate their subject content, and while there is some common ground both books are very relevant. The latter is considerably more advanced and to quote the text on the dust cover 'is geared towards the day-to-day demands of engineers and physicists, etc., etc.'.

Basic instrumentation for engineers and physicists (A. M. P. Brookes. Pergamon, 1968). This book really is basic, but it is easy to read and a good starting point for a student or laboratory worker.

Handbook of applied instrumentation (D. M. Considine and S. D. Ross, eds. McGraw-Hill, 1964). This large volume is as comprehensive as its title suggests. The 17 sections have been compiled by an impressive list of 70 contributors and there is a bibliography of books, papers and standards after each chapter. Inevitably part of the book is concerned with industrial instrumentation, but it is also a good reference tool for the physicist interested in measurement of any parameter.

Handbook of commercial scientific instruments Vol. 1. *Atomic absorption* Vol. 2. *Thermoanalytical techniques* (C. Veillon and W. W. Wendlandt, eds, Dekker, 1972–). These are the first of a multi-volume series devoted to scientific instruments of various types which are commercially available in the United States regardless of country of origin. It is intended that each volume will describe the instrumentation available for a particular field and the material presented will be furnished by the manufacturers with the addition of evaluative comparisons by an author specialised in that field. Subsequent volumes are expected to deal with nuclear magnetic resonance instruments, gas chromatographs, mass spectrometers, X-ray instruments, spectrophotometers and various types of electronic equipment.

Physical laboratory handbook (translated, revised and enlarged by W. Summer from the German original by E. von Angerer and H. Ebert. Pitman, 1966). The German original was first published in 1924 and this is a translation of the 12th edition of 1959. It covers basic laboratory techniques, such as high pressure vacuum, acoustic etc., and also laboratory equipment, such as optical instruments, X-ray tubes, particle detectors and lasers. There are nearly 1000 references.

Techniques générales du laboratoire de physique (J. Surugue, ed. CNRS, 2nd edn; Vol. I, 1955, Vol. II, 1962, Vol. III, 1965). Each of the 21 chapters has been written by an expert in the field and each covers a separate topic, from the essential qualities of an instrument to the methods and equipment for the measurement of the physical quantities and to the use of glass in the laboratory.

Precision measurement and calibration, NBS Special Publication 300 10 vols. 1969–1973 (United States Department of Commerce, National Bureau of Standards). There are so far nine volumes and two more in press. The content belongs more properly to the section on periodicals, since it consists almost entirely of reprints of papers and abstracts, mostly by NBS authors. It is mentioned here because it will be shelved with the books in most libraries and the individual volumes may even be shelved separately in their own subject area, e.g. Temperature, Electricity, Heat, etc.

Transducers

Most books on instrumentation have a chapter on transducers but the following works are entirely about them.

Instrument transducers (H. K. P. Neubert. Clarendon, 1963). A clear and concise presentation of the theoretical and design principles of transducers for high-class instrument work under two main headings, mechanical input and electrical output.

Instrumentation in scientific research: electrical input transducers (K. S. Lion. McGraw-Hill, 1959). This is an earlier work on similar lines which is also frequently cited.

Handbook of transducers for electronic measuring systems (H. N. Norton. Prentice-Hall, 1969). This book is, to quote the dust cover, 'applications oriented'. The first three chapters give the basic principles and the remaining 14 chapters each deal with the measurement of a particular parameter. There are many illustrations, a bibliography and a glossary.

ISA transducer compendium (Instrument Society of America, 2nd edn, in three parts published 1969–72). Its 13 chapters each cover a parameter ranging from the basics, pressure, flow, etc., to radiation and humidity. Each chapter gives general and theoretical information about the measurement of the parameter followed by product information tabulated in such a way that various manufacturers' products can be readily compared. Only products of US manufacture are listed.

Recording—magnetic

There are many books about domestic and commercial tape recording of speech and music, but in general, the physicist will only be concerned with the more sophisticated equipment developed essentially for the recording of high-frequency phenomena and data storage.

Magnetic tape recording (H. G. M. Spratt. 2nd edn, Heywood, 1964). This is one of the best-known on the subject and for many years was the basic textbook, being comprehensive in coverage and compact in presentation.

Magnetic recording in science and industry (C. B. Pear, ed. Reinhold, 1967). Deals with the fundamentals of magnetic tape recording and digital and analogue recording methods, systems and applications. Chapters are contributed by a number of different authors in industry in America, and the book is intended both as an introduction and as a handbook for the experienced.

Digital magnetic tape recording: principles and computer applications (B. B. Bycer. Hayden, 1965). There are several books dealing specifically with digital magnetic tape recording. This one is mentioned as it claims to give comprehensive coverage of the subject including computer applications and is intended to be easily readable. Few mathematical terms and equations are used, and the book is well illustrated.

Recording—photographic

Photography for the scientist (C. E. Engel, ed. Academic Press, 1968). Describes basic photographic materials and processes and practical applications including sections on infra-red recording, ultra-violet and fluorescence recording and closed-circuit television.

Engineering and scientific high speed photography (W. G. Hyzer. Macmillan, 1962). Has chapters on high-speed and low-speed data recording cameras and oscillography and also on film analytical techniques.

Applied photography (D. A. Spencer, ed. Focal Press, 1971). This book is the product of collaboration between three experts, and its 18 chapters include one on each of the following topics: infra-red photography, ultra-violet photography, high-speed photography, instrumentation and recording and the photography of inaccessible objects. There are very many references at the end of each chapter.

Electrical and electronic

As most instrumentation equipment is electrical or electronic, some of the subject matter in this section will have already been dealt with in the previous section on measuring systems. The books selected for mention are just a few of those specifically concerned with the measurement and recording of electrical parameters.

Handbook of electronic instruments and measurement techniques (H. E. Thomas and C. A. Clarke. Prentice-Hall, 1967). Described in the preface as a users' handbook giving information concerning the tools and physical equipment of measurement, namely electronic instruments, and on the procedures and techniques for using these instruments. The book covers such equipment as meters, oscilloscopes, transducers and microwave equipment.

Basic electronic test instruments. Their principles of operation (R. P. Turner. Revised edn, Holt, Rinehart and Winston, 1963). This book covers a similar range of equipment as the one above but with more emphasis on the equipment itself.

Electronics testing and measurement (W. F. Waller, ed. Macmillan, 1972). The foreword claims that 'this book tells the reader how to test the working parameters of electronic components and how to accurately measure electrical values'. The contributors to the 20 chapters work in the electrical industry. The book contains a product guide and a glossary of terms. This is perhaps a book for the physics laboratory technician rather than for the physicist.

All these three books have a bias towards the radio engineer which should not inconvenience the physicist.

Electrical instruments and measurements (W. Kidwell. McGraw-Hill, 1969). This book covers in easily understandable terms the principles of low-frequency and direct current measurements.

Optics and holography

Optical physics (M. Garbuny. Academic Press, 1965). The last 20 pages are about practical detectors of optical radiation.
Laser technology and applications (S. L. Marshall. McGraw-Hill, 1968). Contains a chapter on laser instrumentation.
Laser parameter measurements handbook (H. G. Heard. Wiley, 1968). The preface states that 'it is believed that it treats all the significant laser measurement techniques to date in the areas of beam sampling,

beam parameters, power, energy, gain, wavelength, bandwidth, coherence and frequency stability'. Thirty-seven authors contributed and the bibliography totals more than 650 articles.

Ultra-violet radiation (L. R. Koller. 2nd edn, Wiley, 1965). The last chapter is on detectors of ultra-violet radiation.

Infra-red System Engineering (R. D. Hudson. Wiley, 1969). A very practical book with several chapters on the various types of detectors.

Essentials of modern physics applied to the study of the infra-red (A. Hadni. Pergamon, 1967). A chapter of about 100 pages on infra-red detectors and a very extensive bibliography with a further section on instrumentation in the far infra-red.

The engineering uses of holography. Proceedings of a conference held at the University of Strathclyde September 1968 (E. R. Robertson and J. M. Harvey, eds. CUP, 1970). This substantial volume records the result of collaboration between engineer and physicist which would be useful to either.

Nuclear physics

Nucleonic instrumentation (C. C. H. Washtell and S. G. Hewith. Newnes, 1965). It is stated to be concerned with the essentially electronic equipment designed for measurement and control in the general field of nuclear physics, particularly in research and in the study and handling of radioactive isotopes. The first part of the book is fairly elementary electronics, components and basic circuits, continuing with a description of a number of items of nuclear test instrumentation, pulse counters, rate meters, etc.

Instrumentation and control of nuclear reactors (B. Fozard. Iliffe, 1963). Based on a series of lectures at postgraduate courses and deals with the problems of radiation detection, the various types of detector and how the equipment is applied to reactor instrumentation schemes.

Nuclear reactor instrumentation (in core) (J. F. Boland, prepared under the direction of the American Nuclear Society and the division of Technical Information of the USAEC. Gordon and Breach, 1970). After a discussion of radiation effects, the appropriate equipment is described for the measurement of parameters such as pressure, temperature and neutron and gamma fluid measurements. The material is well presented and includes extensive bibliographies.

Plasma physics

Of the several books on plasma diagnostics the following two have been selected as having some material on instrumentation:

Plasma diagnostic techniques (R. H. Huddlestone and S. L. Leonard, eds. Academic Press, 1965). Three chapters are devoted to spectroscopic diagnostics in the optical and ultra-violet, X-ray and far infra-red portions of the electromagnetic spectrum, followed by optical interferometry, microwave techniques and particle measurements, in all cases with emphasis on instrumentation.

Plasma diagnostics (W. Lochte-Holtgreven, ed. North-Holland, 1968). This book covers a much wider field but does include a substantial part on the topics mentioned in the previous book and also on the application of the laser in plasma diagnostics.

Temperature

Since temperature is such a basic measurement, there is plenty of material from which to choose, and the physicist should have few selection problems. The following are representative works.

Bibliography of temperature measurement January 1953 to December 1969. NBS Special Publication 373 (US Department of Commerce, National Bureau of Standards, 1972). This work is divided chronologically into four sections and each section is sub-divided into the same 11 subject headings. The material is collected from two general sources, i.e. scientific and technical journals and reports of investigations sponsored or conducted by various governmental agencies. There are in all over 4000 references, 1000 of which are in the section covering the period January 1966 to December 1969.

Thermocouple temperature measurement (P. A. Kinzie. Wiley/Interscience, 1973). This book summarises information on both well-known and little-used thermocouples, and is intended to provide a convenient source of information for the study of unconventional requirements and their solutions. The appendix includes an extensive summary of thermocouple types and characteristics, and 27 pages of references, including patents.

Physicochemical measurements at high temperature (J. O'M. Bockris *et al.*, eds. Butterworths, 1969). The authors of each of the 15 chapters are from several different countries, but good uniformity has been achieved and the book is compact and well illustrated, and each chapter has many references. The first four chapters contain general information on high-temperature investigation, and each of the remaining 11 chapters discusses techniques for investigating a given type of property, e.g. surface tension, ultrasonic velocity.

Temperature measurements in seeded air and nitrogen plasmas (H. N. Olsen *et al.* Management Information Services, Detroit, 1970). This report of more than 150 pages describes the methods and instrumentation used at the Arnold Engineering Development Centre for the measurement of temperatures in the range 2500 to 5000 K.

Cryogenic laboratory equipment (A. J. Croft. Plenum, 1970). The preface states that the book is for laboratory workers who for one reason or another have a need to cool something down to temperatures below that of liquid nitrogen. It is therefore a very practical book. It includes a chapter on simple instrumentation for the measurement of basic parameters as well as potentially useful information on materials and jointing methods.

The following books also each have a chapter on instrumentation for low-temperature measurement:

Cryogenic technology (Robert W. Vance, ed. Wiley, 1963)

Cryogenic engineering (J. H. Bell. Prentice-Hall, 1963)

ABSTRACTS JOURNALS AND PERIODICALS

There are no abstracts journals entirely devoted to instrumentation, but most abstracts journals in the field of physics have a section on instrumentation and testing, and the following notes will indicate the most suitable starting point according to the particular branch of physics or the preferred language.

The *Science Citation Index* and its associated *Permuterm* are not discussed here, as they have been described elsewhere.

On the other hand, there are many periodicals on instruments and instrumentation. Some of these periodicals are fully abstracted and the physicist may be guided to the ones most suited to his needs by a study of the abstract journals. Other periodicals have only the occasional article abstracted and there will be much of current interest in the smaller articles, letters and advertisements. Periodicals mostly concerned with the engineering aspect of the equipment or with industrial instrumentation have not been included; neither have any of the periodicals on applied physics which do contain the occasional but significant article on measurement and its associated equipment.

Abstracts journals

Physics Abstracts, Science Abstracts, Series A and *Electrical and Electronics Abstracts, Science Abstracts Series B.* These are the two most significant abstract journals in the field of instrumentation in physics and have been described fully in an earlier chapter. Both are produced by INSPEC and there is a small overlap of articles equally relevant to both journals. The revised subject arrangements introduced at the beginning of 1973 provide a number of convenient subheadings under the main headings of Physical Instrumentation and Experimental Techniques and Instrumentation and Special Applications, respectively.

Current Papers in Physics and *Current Papers in Electrical and Electronics Engineering* are also produced by INSPEC and have the same subject arrangements as the two previous journals. Only the article reference is given and they are intended to be used for current awareness.

Metron (SIRA Institute, Chislehurst). Monthly. This is also produced by INSPEC and is entirely devoted to measurement, control and instrumentation. It is not strictly an abstract journal, as the articles are listed and not abstracted, apart from a single line of explanation, but they are grouped by aspect and by application and are indexed. In addition, there is each month a technical note on a particular subject and also an information review. As an example of the relevance of the latter, the January 1975 entry was about thermal measurements and the May 1975 entry about mechanical parameter measurements.

Bulletin Signalétique (CNRS). This is the principal French abstract journal and is published in separate sections (50 at the time of writing), each concerned with a particular subject. The eight sections concerned with physics are numbered between 120 and 165. References on measurement and instrumentation will generally be found either as a sub-division of the subject heading 'Métrologie' or in an adjacent section.

Referativnyi Zhurnal 32, Metrologiya i izmeritel'naya tekhnika (Akad. Nauk). Monthly. This section of the Russian abstract journal covers metrology, instrumentation and measurement techniques. There are about 15 subject headings, each concerned with a physical parameter or grouped associated parameters. At present there is an annual subject index but no author index.

Physikalische Berichte (Deutsche Physikalische Gesellschaft).

Monthly. Has important sections on measurement and instrumentation and a very wide coverage of periodicals.

Physics Express. This journal is published in America and 'culls' about 100 Russian periodicals. Some of the entries are merely a translation of the title, while others are a full abstract, sometimes with diagrams. There is a section entitled 'Instrumentation'.

Nuclear Science Abstracts, published semi-monthly by the USAEC Office of Information Services, claims to provide the only comprehensive abstracting and indexing coverage of the international nuclear science literature. It has a section on instrumentation which is mostly concerned with radiation detection and effects. Among the items abstracted are US atomic energy reports, patents and conference papers.

The following are representative of abstract journals covering a narrow subject field in physics but including some aspects of measurement.

Surface Wave Abstracts (Multi-Science Publishing Co.). Quarterly. Has section on measurement and recording.

Acoustics Abstracts (Multi-Science Publishing Co.). Covers measurement and recording of sound at various frequencies, e.g. ultrasonic and under various conditions, e.g. underwater.

Periodicals

Journal of Physics E: Scientific Instruments (Institute of Physics). Published monthly with annual subject and author indexes. Consists of one or two main articles and about 25 shorter contributions under the main headings of Apparatus and Techniques and Research Papers. There are also a few letters to the editor and reviews of books and manufacturers' literature.

Review of Scientific Instruments (American Institute of Physics). Also published monthly with annual subject and author indexes. Consists of about 20 articles and a number of single-page Notes, lists of books received and some book reviews, and also descriptions of new instruments and new materials and components.

Soviet Instrumentation and Control Journal (Robert Maxwell, Oxford). A cover to cover translation of the Russian journal *Pribory i Sistemy Upravleniya* appearing about 12–18 months later. Some industrial instrumentation content but all of general interest to the physicist.

Instruments and Experimental Techniques (Plenum). A cover-to-cover translation under the editorial direction of the Instrument Society of America of the Russian journal *Pribory i Tekhnika Eksperimenta* appearing about 6 months after the publication of the original Russian issue. The periodical is bi-monthly and contains upward of 50 fairly short articles, all very much in the field of physical instrumentation. There is a section dealing with the practical application of some instruments and a review of new instruments.

IEEE Transactions on Instrumentation and Measurement (Institute of Electrical and Electronics Engineers). Quarterly. Issue content is variable, mostly comprising up to a dozen papers and a smaller number of short contributions, though some issues are very much bigger and may contain conference proceedings. As might be expected all the equipment and methods described are electrical or electronic.

Instruments and Control Systems (Chilton). Monthly. Free in US and Canada to qualified individuals. Much of the subject matter is concerned with industrial instrumentation, but it may be of interest to the physicist, because it is one of the few periodicals mentioned which includes advertisements of readily available equipment and components.

Control and Instrumentation (Morgan Grampian). Monthly. Free in UK to qualified persons. (Content similar to that of previous reference.)

Metrologia (International Journal of Scientific Metrology). Published four times a year under the auspices of the International Committee of Weights and Measures (Springer). This periodical 'invites for publication articles that report the results of original researches directed towards the significant improvement of fundamental measurements in any field of physics'. Most of the articles are in English and authorship is world-wide, mostly by physicists in standards or physics laboratories.

Nuclear Instruments and Methods. A journal on accelerators, instrumentation and techniques in nuclear physics. (North-Holland.) Semi-monthly. The present editor is at the Institute of Physics, University of Uppsala and the large editorial board is international. A very substantial publication, mostly in English. Started in 1957 and now producing six volumes per year with individual author and subject indexes and a cumulative index every 10 years.

IEEE Transactions on Nuclear Science (Institute of Electrical and Electronics Engineers). Bi-monthly. Content of individual issues variable from a few articles on nuclear science or nuclear instrumentation to proceedings of conferences in the subject field.

Advances in Instrumentation (Instrument Society of America). This is now an annual four-part publication recording the proceedings of the ISA annual conference. These present an up-to-date report on the state of the art in instrumentation. Each year a different theme is selected, and although much is about industrial instrumentation, the series is well worth scanning for the papers on new measurement techniques.

Instrumentation Index (Instrument Society of America). Quarterly. This publication has been designed as a current-awareness tool as well as a means of searching for articles by author and by publication. It covers all ISA books and journals. The first part is a KWIC index and the second part a list of titles in each publication and an author index.

ISA Transactions (Instrument Society of America). Quarterly. Although much of the subject matter is concerned with industrial instrumentation, in most issues there are several articles on unusual and wide-ranging aspects of measurement. The annual subject index is clear and concise.

Messtechnik (Vieweg). Monthly. The main text is in German with summaries of the principal articles in English. There are four or five articles on testing and measurement and associated equipment, each with a useful bibliography, largely from German sources, and a section on new products.

Laboratory Equipment Digest (Gerard Mann, London). Monthly. A substantial publication with a few multi-page articles, book reviews, news, exhibition reports, forthcoming events, but with much of the space occupied by short paragraphs on new equipment and advertisements. Full of useful current-awareness information for the physicist.

BUYERS' GUIDES AND EXHIBITION CATALOGUES

Guide to Japan's measuring instruments/controls 1971 (Dempa Publications, Tokyo), in English. There are three main sections: catalogue, reference and directory. It is particularly useful for its information on Japanese measuring instruments, oscillographs, tape recorders, etc., and components, counters, relays, solenoids, etc.

British instruments directory (United Trade Press in association with the Scientific Instrument Manufacturers' Association of Great Britain). Annual. Provides an alphabetical list of instrument and component manufacturers, a classified list of instruments and components and a number of reference sheets describing equipment.

IEA instruments, electronics, automation (Morgan Grampian, publishers of *Control and Instrumentation Electronic Engineering*).

Annual. Lists trade names and manufacturers' addresses, and provides a buyers' guide and some advertisements.

EEM electronic engineers master (United Technical Publications, New York). Annual. Three-volume directory covering test instruments and systems and components.

Laboratory equipment directory and buyer's guide (Mann). In addition to classified buyers' guide and manufacturers' addresses, also gives trade names and UK agents of overseas organisations with products in the laboratory equipment field.

Japan EBG electronic buyers' guide (Dempa Publications, Tokyo). Comprehensive directory including review of Japan's electronic industries and foreign firms' representatives as well as conventional product and trade name information.

16

Computer applications

C. D. M. Johnston

This chapter is not intended to be a general guide to the literature of computers. Instead it is hoped that it will provide access to that part of the literature likely to be of interest to physicists who wish to make use of computers in their work. The literature concerned is exceedingly diffuse. It ranges from introductory texts on computers and computing, through books of various levels on programming, to specialised papers on applications in different branches of physics. One of the problems in this field is that while there are a few specialist journals, the majority of information is scattered throughout the literature of physics and, to a lesser extent, throughout the computer literature as well. It is also a subject which is relatively new, but which is expanding very rapidly; consequently, some books published 10 years ago seem already completely out-of-date.

For those who wish to go more thoroughly into the literature of computers than there is space for in this chapter, the following book is strongly recommended: *A guide to computer literature*, by A. Pritchard (2nd edn, Bingley, 1972). This covers the subject very thoroughly and includes chapters on reports, trade literature, conferences, bibliographies, annual reviews and industry statistics, as well as on the more straight forward aspects, such as books, periodicals and abstract journals.

For more information on books, one can refer to *International computer bibliography* (National Computing Centre, Manchester, in association with Stichting Het Nederlands Studiecentrum voor Informatica, Amsterdam. Vols. 1 and 2, 1968, 1971). This is an annotated list of books (including some pamphlets and reports) on all aspects of computers, arranged in a classified order. It includes indexes of keywords and authors.

In the next part of this chapter, details of a number of books are given, mainly ones on the use of computers rather than on their construction and circuitry (the 'hardware'). Then follow sections on abstract journals and on periodicals, with a final section on trade literature and trade directories.

BOOKS

General and introductory books

A book which will be of particular interest to physicists is *Computers and their role in the physical sciences,* edited by S. Fernbach and A. H. Taub (Gordon and Breach, 1970). It is written for the 'non-expert reader' by a series of authors. It starts with a historical introduction and this is followed by a brief description of computer hardware, which omits some essential parts, such as input/output equipment and high-capacity data storage devices. There is a short section on software and then follows a series of chapters (of variable quality) on the impact of computers on mathematics and the various branches of physics.

A useful general textbook is *Digital computer user's handbook*, edited by M. Klerer and G. A. Korn (McGraw-Hill, 1967). It is a 'McGraw-Hill Handbook' and is as comprehensive as these handbooks normally are. Including only a very brief description of computers themselves, it concentrates largely on the use of computers for numerical analysis and statistical methods. Each chapter contains a number of references for further reading.

Another from the same stable is *Computer handbook* edited by H. D. Huskey and G. A. Korn (McGraw-Hill, 1962). Although rather old now, it can still be recommended, particularly for information on analogue computers. The emphasis is on the hardware rather than on applications.

Three books which could serve as introductions to computing for someone who has little or no prior knowledge are: *A first course in computing and numerical methods*, by J. A. Jacquez (Addison-Wesley, 1970); *Introduction to Fortran programming*, by H. M. Liddell and A. J. Powell (Harrap, 1971); and *Introduction to computing*, by T. E. Hull (Prentice-Hall, 1966). The first book starts with basic concepts, such as how numbers are represented and handled in a computer, and briefly describes programming in MAD, Fortran and Algol. The rest of the book deals with the way in which a computer can be used for various mathematical techniques. The second book is shorter and is restricted to Fortran, but is illustrated by reference to some scientific applications. The third book also deals with Fortran and is written at a very basic level, but contains a useful bibliography.

Both *Minicomputers for engineers and scientists*, by G. A. Korn (McGraw-Hill, 1973), and *An introduction to on-line computers*, by W. W. Black (Gordon and Breach, 1971), are more specialised than those mentioned previously, but both are relevant to applications in physics. The first one, particularly, is written from a very practical viewpoint and includes an extensive bibliography.

Many applications of computers in physics involve problems of data collection, prior to data processing; the following book is therefore likely to be of interest (most of the papers cover the instrumentation as well as the data processing): *The collection and processing of field data, a CSIRO symposium*, edited by E. F. Bradley and O. T. Denmead (Interscience, 1967).

As well as problems of data collection, physicists are often involved in more general problems of input to and output from computers, and may wish to use special-purpose peripheral equipment. The next books should help in this: *Digital interface design*, by D. Zizzos and F. G. Duncan (OUP, 1973) and *Minicomputer interfacing* (Proc. Symp. at N. London Polytechnic), edited by Y. Paker *et al.* (Miniconsult, 1973).

Some of the publications of the National Computing Centre (Manchester, England) are worth considering in the context of physics applications, although many are more for the business user than for the scientist. Some are written for the informed layman— for instance, the series *Computers and the manager*. Others are authoritative guides produced to encourage good practice. Examples of the latter are: *Programming standards*, Vol 1: *Documentation*, Vol 2: *Techniques* (National Computing Centre, 1972) and *Standard FORTRAN programming manual* (2nd edn, National Computing Centre, 1973).

Applications of computers in physics

In addition to *Computers and their role in the physical sciences* (Fernbach and Taub), another useful general book, which deals more with the mathematical aspects, is *Computational physics*, by D. Potter (Wiley, 1973). The author aims to provide a reference work for those applying computers in particle theory, fluid dynamics, etc.

An extremely valuable series is published under the title *Methods in computational physics, advances in research and applications*

(Academic Press). Each volume deals with a particular branch of physics and those published so far are:

Vol. 1 *Statistical physics* (1963)
    2 *Quantum mechanics* (1963)
    3 *Fundamental methods in hydrodynamics* (1964)
    4 *Applications in hydrodynamics* (1965)
    5 *Nuclear particle kinematics* (1966)
    6 *Nuclear physics* (1966)
    7 *Astrophysics* (1967)
    8 *Energy bands of solids* (1968)
    9 *Plasma physics* (1970)
   10 *Atomic and molecular scattering* (1971)
   11 *Seismology: surface waves and earth oscillations* (1972)
   12 *Seismology: body waves and sources* (1972)
   13 *Geophysics* (1973)

The proceedings of a number of conferences are a useful source of information in this field. For example: *Computational physics. Proceedings of the conference at Culham in 1969. 2 vols.* UKAEA London, 1969. Available through HMSO. The first volume consists of nine review papers, starting with one on 'Methods of computational physics', followed by eight on the use of computers in different branches of physics. Volume II consists of 53 papers on more specific topics but grouped under the same eight headings as the review papers.

Computational physics. A digest of the proceedings of the 2nd conference. Institute of Physics and The Physical Society, 1970. This volume contains the digests of 29 papers covering a wide range of branches of physics. Although called 'digests', they are not unduly abridged, and references are included.

Computing as a language of physics. Lectures presented at the International Seminar, Trieste, 1971. International Atomic Energy Agency, Vienna, 1972. The seminar includes three general review papers on the use of computers in physics and in nuclear science and also a very valuable paper 'Comparative survey of programming languages', by J. A. Cambell. This paper ends with a list of references, which include introductory references to each of the languages discussed.

The impact of computers on physics, the first European conference on computational physics, 1972. Invited papers were published as a supplement to Vol. 3 (1972) of *Computer Physics Communications* (Netherlands). Contributed papers appear in various issues of Vols. 4 and 5 (1972–73). Most of the invited papers review the current state of the art in different branches of physics.

Finally, the following books are rather more specialised but both include extensive lists of references: *Computing methods in reactor physics*, edited by H. Greenspan *et al.* (Gordon and Breach, 1968), and *Computational fluid dynamics*, by P. J. Roache (Hermosa, 1972).

Programming digital computers

There must have been more books written on computer programming in the last few years than on any other subject. From these a few books of particular interest to scientists have been selected. Most of the basic textbooks and handbooks mentioned above deal with programming to some extent, but the books listed here go deeper into the subject.

Written at a fairly elementary level but useful none the less in discussing the main features and advantages of each language is *Computer languages: a practical guide to the chief programming languages*, by P. C. Sanderson (Newnes-Butterworth, 1970). A more advanced book, written for research workers in science and engineering, discusses the use of Fortran, PL/1 and Algol, and contains a list of references: *Computer programming and computer systems*, by A. Hassitt (Academic Press, 1967). The following book covers the subject fairly thoroughly, includes a very good bibliography and recommends references for each language discussed: *Programming languages: history and fundamentals*, by J. E. Sammet (Prentice-Hall, 1969).

The basic subject of the following series is described as 'non-numerical analysis'. Of the three volumes so far published, the first two could certainly be of interest to physicists.

The art of computer programming. D. E. Knuth. Addison-Wesley, 1968–
   Vol. 1: *Fundamental algorithms* (1973)
   Vol. 2: *Seminumerical algorithms* (1969)
   Vol. 3: *Sorting and searching* (1973)

Particularly for users of Algol, a series of standard texts has started to appear:

Handbook for automatic computation. F. L. Bauer *et al.* (eds.). Springer, 1967–
   Vol. Ia: *Description of Algol 60* (1967)
   Vol. Ib: *Translation of Algol 60* (1967)
   Vol. II: *Linear algebra* (1971)

The aim of the handbook is 'to supply a selection of tested algorithms for the solution of standard problems in numerical analysis'. Future volumes are intended to cover: linear algebra; functional equations, in particular differential equations and integral equations; methods of approximation; evaluation of special functions.

At a less advanced level, good introductions to Algol are: *An introduction to Algol programming*, by R. Wooldridge and J. F. Ractliffe (3rd edn, English Universities, 1968), and *Introduction to Algol*, by R. Baumann *et al.* (Prentice-Hall, 1964).

Both the following books on PL/1 are written for the scientist or engineer (the first is rather more elementary than the second, which assumes some programming experience): *PL/1 programming in technological applications*, by G. F. Groner (Wiley/Interscience, 1971), and *PL/1 for scientific programmers*, by C. T. Fike (Prentice-Hall, 1970).

For basic and APL the following books are suggested: *Programming time-shared computers in BASIC*, by E. H. Barnett (Wiley/Interscience, 1972), and *APL programming and computer techniques*, by H. Katzan (Van Nostrand Reinhold, 1970).

For someone intending to use an IBM 360, who already has some programming experience, a particularly useful book should be: *Programming the IBM System/360*, by R. F. Steinhart and S. V. Pollack (Holt Rinehart and Winston, 1970). It describes the use of various languages such as Assembler Language, Fortran IV, Cobol, PL/1 and Algol, and would be helpful in selecting the most suitable language for a particular application.

Analogue and analogue/hybrid computers

Analogue and hybrid computers are used more in scientific research than in commercial work. Many of the books are therefore aimed at the scientist or engineer, in contrast to books on digital computers, so many of which are written for the business user.

Two good introductory textbooks are: *The design and use of electronic analogue computers*, by C. P. Gilbert (Chapman and Hall, 1964), and *Analog computer programming*, by M. G. Rekoff (Merrill, 1967). The latter deals mainly with the use of the computer, whereas the first book also covers the design of the computer itself.

A more advanced book, translated from the Czech, is mainly on the use of analogue computers and has little on their design and circuitry: *Computation by electronic analogue computers*, by V. Borsky and J. Matyáš (Iliffe, 1968).

A very comprehensive book, mainly on the hardware, is: *Electronic analog and hybrid computers*, by G. A. Korn and T. M. Korn (2nd edn, McGraw-Hill, 1972). The new edition of this excellent book deals mainly with analogue/hybrid computers rather than with the pure analogue machine. Nevertheless it covers all the basic components of an analogue computer and the effect of their accuracy on computer performance. Input and output equipment is described, including the interface with the digital computer. A chapter on computing techniques includes some of the new mathematical techniques opened up by analogue/hybrid computation and describes the use of the iterative differential analyser. Lists of references at the end of each chapter combine to form a bibliography of 730 items, with an author index providing alphabetical access to it.

More on the software side is: *Analog and analog-hybrid computer programming*, by A. Hausner (Prentice-Hall, 1971). It concentrates mainly on the use of the computer for solving classes of mathematical problems. Many references are given at the end of each chapter.

An excellent book for the physicist intending to use a hybrid computer is: *Hybrid computation*, by G. A. Bekey and W. J. Karplus (Wiley, 1968). The second half includes some applications in physics, e.g. the solution of field problems. Many references are listed.

Direct analog computers, by V. Paschkis and F. L. Ryder (Interscience, 1968) covers a type of analogue computer which is rather different from those described in the other books listed. It describes a number of applications in physics, such as heat transfer and fluid flow.

Simulation and modelling

One of the important ways in which computers are used by physicists is in simulation, and for this purpose a digital, analogue or hybrid computer may be used. Some of the books on analogue and hybrid computers which have already been mentioned include some treatment of simulation but the following are specifically on this technique.

Computer modelling and simulation, by F. F. Martin (Wiley, 1968) and *Simulation. The dynamic modelling of ideas and systems with computers,* edited by J. McLeod (McGraw-Hill, 1968) are introductory and not specifically on applications in physics. The first includes a good bibliography; the latter mainly consists of selected reprints from the journal *Simulation*.

Digital simulation of continuous systems, by Y. Chu *et al.* (McGraw-Hill, 1969). This book was written to accompany an undergraduate course for engineers and scientists and its emphasis is on programming.

Random-process simulation and measurements, by G. A. Korn (McGraw-Hill, 1966) is concerned with a specialised facet of simulation but one that has applications in some areas of physics. The first two chapters provide a concise review of random-process mathematics and of analogue/hybrid computation and the author goes on to describe techniques for random-process studies, using analogue, digital and hybrid computers. Instrumentation and practical procedures are described, as well as the underlying theory.

Analogue-to-digital conversion

In using digital computers for physics research, there is often a need to convert the analogue signals from transducers into digital form, so that they can either be fed straight to a digital computer or recorded for processing later. The following books are of interest:

Electronic analog/digital conversions. H. Schmid. Van Nostrand Reinhold, 1970. This book does not include mechanical analogue-to-digital converters, but apart from that it covers the field thoroughly and many references are given.

Analog-to-digital/digital-to-analog conversion techniques. D. F. Hoeschele. Wiley, 1968. This is not quite as detailed on electronic devices as the previous book, but includes shaft position encoders.

Notes on analog-digital conversion techniques. A. K. Susskind (ed.) Technology Press of MIT and Wiley and Chapman and Hall, 1957. Rather old, but this is regarded as a classic and goes into the subject in considerable detail.

Computer graphics

For many applications in physics, an output in graphical form can sometimes be more useful than a numerical print-out. *Computer graphics. Techniques and applications*, edited by R. D. Parslow and R. W. Prowse (Plenum, 1969) is based on the papers presented at a symposium held at Brunel University. It starts with a general review of the technique and applications. There are papers on various types of application and two in particular are worth mentioning here. 'Graphical output in a research establishment' describes applications at UKAEA Culham Laboratory and 'High energy physics applications' describes work at CERN. There is a bibliography at the end of the book.

Mathematical methods and numerical computation

Very many books have been published on these subjects and it would not be appropriate here to attempt to review the whole field. Instead a few books have been selected—the first group are mainly of the comprehensive handbook type and these are followed by some more specialist books. Some of the books listed above are also relevant in this context, particularly Klerer and Korn's *Digital computer user's handbook*.

Mathematical handbook for scientists and engineers. G. A. Korn and T. M. Korn. 2nd edition. McGraw-Hill. 1968. Although not covering the use of computers as such, this book is a comprehensive reference book covering those mathematical techniques likely to be needed by scientists or engineers. The most important formulae and definitions are collected in tables which are useful for quick reference. Each chapter contains a short list of references.

Handbook of numerical methods and applications. L. G. Kelly, Addison-Wesley, 1967. The first half of this book describes the basic numerical methods used by scientists and engineers, while the rest of the book covers a variety of special topics. These include harmonic analysis, sampled data and digital filtering, numerical solution of vibration problems, etc. The book is well served with references in each chapter and a more extensive bibliography at the end.

Survey of applicable mathematics. K. Rektorys *et al*. Iliffe, 1969. This is a very large book (over 1000 pages), translated from Czech, and covering almost every branch of mathematics that a physicist or engineer might need. The contents are clearly presented, with much tabular information, including (so it claims) the most complete table of solutions of differential equations published in English. There is a list of nearly 500 references.

The next four books are considerably smaller than those mentioned earlier and have been selected because they are essentially practical books written for scientists or engineers. In the first one, each chapter contains a lengthy bibliography.

Applied numerical methods. B. Carnahan *et al*. Wiley, 1969

Computer applications of numerical methods. S. S. Kuo. Addison-Wesley, 1972

Computing methods for scientists and engineers. L. Fox and D. F. Mayers. Clarendon, 1968

Numerical methods for scientists and engineers. R. W. Hamming. 2nd edition. McGraw-Hill, 1973.

Each of the remaining books listed in this section deals with a rather more specialised aspect of the subject:

A handbook of numerical matrix inversion and solution of linear equations. J. R. Westlake. Wiley, 1968

Computer evaluation of mathematical functions. C. T. Fike. Prentice-Hall, 1968

Fitting equations to data. C. Daniel and F. S. Wood. Wiley/Interscience, 1971

Handbook of mathematical functions, with formulas, graphs and mathematical tables. M. Abramowitz and I. A. Stegun (eds.). National Bureau of Standards, 1970

Random data: analysis and measurement procedures. J. S. Bendat and A. G. Piersol. Wiley/Interscience, 1971

Encyclopaedias and dictionaries

A small but very useful book to have at one's elbow is: *A dictionary of computers*, by A. Chandor *et al.* (Penguin, 1970). In some cases it goes well beyond a definition of a term and provides a brief introduction to some basic topics, together with suggestions for further reading.

Two rather larger books are: *Condensed computer encyclopedia*, by P. B. Jordain and M. Breslau (McGraw-Hill, 1969), and *Computer dictionary and handbook*, by C. J. Sippl (Foulsham, 1967). The first half of the latter is an extensive glossary of computer terminology and jargon (most important in this field) and the second half contains an assortment of useful information including acronyms and abbreviations.

Another glossary is: *Glossary of computing terminology*, by C. L. Meek (CCM Information Corporation, 1972). This includes lists of commercially available equipment, arranged both numerically (by type number) and alphabetically (by trade name), with brief descriptions.

ABSTRACTS JOURNALS

Computer and Control Abstracts (Institution of Electrical Engineers). This is probably the most fruitful source of information on the applications of computers in physics. Section 8.814 is on 'Applications in physics' (before 1973, 88.12) and section 8.200 on 'Numerical analysis' is also relevant (before 1973, 82.00).

The companion journal *Physics Abstracts* should also be consulted; section 1.180 covers 'Numerical methods and computational physics'. *Physics Abstracts* has been described in detail in Chapter 5. *Computer and Control Abstracts* has a similar format and similar indexes.

Though not containing abstracts, it is worth mentioning *Current Papers in Physics* and *Current Papers on Computers and Control*. Also published by the Institution of Electrical Engineers, these provide more rapid information of new publications, classified in the same way as the two abstract journals already mentioned.

Computer and Information Systems (Cambridge Scientific Abstracts, Riverdale, Maryland, USA). This abstract journal has a good section on 'Computers in the physical sciences and engineering'. There is a subject index in each issue, with annual cumulations.

Computer Abstracts (Technical Information Co., Jersey). This journal principally covers scientific applications and programming, and so is well suited to the physicist. It gives extensive coverage to numerical methods and algorithms. There is a section on 'Physics applications' and another on 'Nuclear applications'. Each issue includes a subject index and these are cumulated annually.

Computing Reviews (Association for Computing Machinery, New York). This 'aims to furnish computer oriented persons in mathematics, engineering, the natural and social sciences, the humanities, and other fields with critical information about all current publications in any area of the computing sciences'. Section 3.17 covers 'Applications in physics; nuclear sciences'. Annual KWIC indexes are provided.

Computer Program Abstracts (National Aeronautics and Space Administration, Washington). This is a specialised abstract journal, confining itself to computer programs developed by or for NASA and the US Department of Defense and available for sale. The abstracts are arranged by subject, with sections on several branches of physics. There is a subject index in each issue.

Proceedings of the International Association for Analog Computation, also entitled *Annales de l'Association Internationale pour le Calcul Analogique* (Presses Académiques Européennes, Brussels). Not primarily an abstract journal, but it includes a useful section of abstracts and covers hybrid as well as analogue computers.

In addition to *Physics Abstracts*, there are some other general abstract journals which may be useful. These have been described in more detail in Chapter 5 and so are only mentioned briefly here.

Bulletin Signalétique (CNRS). Section 110 covers (amongst other topics) the application of computers in physics and also in nuclear energy.

Physikalische Berichte (Deutsche Physikalische Gesellschaft, Braun-schweig). There is a section on mathematical physics.

Nuclear Science Abstracts (US Atomic Energy Commission). There is a small section on 'Mathematics and computers' some of which is relevant.

Scientific and Technical Aerospace Reports (NASA Scientific and Technical Information Office, Washington). This has a small section on computers, but references elsewhere may be indexed under 'Computer' in the subject index.

Referativnyĭ Zhurnal (VINITI, Moscow). The two series of this Russian abstract journal which are relevant here are *Avtomatika, Telemekhanika i Vychislitel'naya Tekhnika* and *Kibernetika*. The former series is largely on the hardware and engineering applications, although it does cover analogue computers and their use in physical modelling. *Kibernetika*, on the other hand, includes programming and also applications in physics and mathematics.

Cybernetics Abstracts (Scientific Information Consultants Ltd, London). This is a selective translation of *Referativnyĭ Zhurnal, Kibernetika*.

In the field of mathematical methods and numerical computation, the following abstract journals are recommended:

Mathematical Reviews (American Mathematical Society, Providence). This has sections on numerical methods and on computing machines, as well as on the use of mathematics in the various branches of physics.

Index of Mathematical Papers (American Mathematical Society, Providence). This is an index of papers, without abstracts. The subject index in each issue includes similar sections to those mentioned above under *Mathematical Reviews*.

PERIODICALS

Computer Physics Communications (North-Holland). This is the main European journal in the field and is edited from Queens University, Belfast. It started in 1970 and for the first 2 years the emphasis was on papers describing programs held and distributed by the CPC Program Library in Belfast. From 1972 the scope was extended to include papers in the general area of computational physics.

Journal of Computational Physics (Academic Press). This is the principal journal in the USA on the subject. Its stated policy is to publish 'articles concerning techniques developed in the solution of data handling problems and mathematical equations, both arising in the description of physical phenomena'. Not all the papers concern computers and some include descriptions of computations which can be done without the aid of a computer.

Methods in Computational Physics (Academic Press). This is largely edited by the same team as the *Journal of Computational Physics* and is a review series in which each volume deals with one aspect of the subject. It is described in more detail above (p. 260).

Computers and Structures (Pergamon). Although primarily of interest to the engineer, it contains some papers of potential interest to the physicist. It mainly covers 'the applications of computers (digital, analog and hybrid) and computer programs to the solution of scientific and engineering problems related to hydrospace, aerospace and terrestrial structures'. It also tries to cover applications in fluid mechanics, geophysics and materials science, but recent issues have contained little on these topics.

Computers and Fluids (Pergamon). This only started in 1973 and so it remains to be seen how it develops. It aims to be 'multidisciplinary, interpreting the term "fluid" in a broad sense'.

CAMAC Bulletin (Esone Committee, Luxembourg). CAMAC is the designation of rules for the design and use of modular electronic data-handling equipment. It offers a standard scheme for interfacing computers to data transducers and actuators in on-line systems. Although originally conceived for nuclear research, it is now being applied in other fields. The *Bulletin* provides information on CAMAC activities, commercially available equipment, applications, etc. Recent issues have contained a bibliography on the subject.

The periodicals so far listed are the core journals in so far as computer applications in physics are concerned. The following list consists of the more academic journals covering computers or computing in general, and most have some papers of potential interest to physicists using computers.

Advances in Computers (Academic Press). This is a review series, with the emphasis on applications and software rather than on hardware.

Bit (Nordisk Tidskrift for Informationsbehandling, Copenhagen). This includes lists of new books, news of conferences, etc., as well as research level papers. It is entirely in English.

Communications of the ACM (Association for Computing Machinery, Baltimore); *Journal of the Association for Computing Machinery* (Baltimore, USA). Of the two from the ACM, the *Journal* is the more academic, while the *Communications* tends to have shorter papers, general news and coming events.

Computer Journal (British Computer Society, London). As well as papers (both original and review), each issue normally includes a number of book reviews.

Computing (Springer). It includes book reviews and is partly in English.

IEEE Transactions on Computers (Institute of Electrical and Electronics Engineers)

Proceedings of the International Association for Analog Computation (Brussels)

Revue Française d'Automatique, Informatique, Recherche Opérationnelle (Dunod)

Siam Journal on Computing (Society for Industrial and Applied Mathematics, Philadelphia)

Simulation (Society for Computer Simulation, La Jolla, California)

Software Practice and Experience (Wiley/Interscience)

Software World (A. P. Publications, London)

For general news of the industry, announcements of new equipment, forthcoming conferences, etc., the following are suggested.

Computer Weekly (IPC Electrical-Electronic Press, London)

Computerworld (Boston, Mass)

These two weeklies, one from each side of the Atlantic, are similar—both are in newspaper format.

Computing (Haymarket Press, for the British Computer Society)

Computing Report in Science and Engineering (IBM, White Plains, New York). Mainly news of how IBM computers are being used, but it is also a valuable source of information on the large number of publications produced by IBM.

Datamation (Technical Publishing Co, Barrington, Illinois)

NCC Interface (National Computing Centre, Manchester). Useful chiefly for news of activities of the NCC and of its publications.

For numerical analysis, mathematical computation, mathematical physics, etc., some of the academic type of periodical already listed contain papers on these topics. The following will also be of interest:

Journal of Mathematical Physics (AIP)

Mathematical Programming (North-Holland)

Mathematics of Computation (American Mathematical Society, Providence)

Numerische Mathematik (Springer). Most papers are in English.
SIAM Journal on Applied Mathematics (Society for Industrial and Applied Mathematics, Philadelphia)
SIAM Journal on Mathematical Analysis
SIAM Journal on Numerical Analysis
SIAM Review
USSR Computational Mathematics and Mathematical Physics (Pergamon). Cover-to-cover translation of *Zhurnal Vychislitel'noĭ Matematiki i Matematicheskoĭ Fiziki*.

TRADE LITERATURE, TRADE DIRECTORIES, ETC.

As a source of information on computers and their applications, trade literature can be invaluable. This includes not just glossy brochures describing the main features of equipment on offer, but manuals describing in detail individual machines, software and applications. All the major manufacturers produce a great deal of this type of literature, and it can be useful not only in deciding what computer to buy or rent but also in making maximum use of it once it is installed. Bibliographical control of it is not easy but most companies have lists available.

Apart from the information available from individual manufacturers, the most comprehensive source of information on commercially available equipment is Auerbach Publishers Inc. This company produces several series of loose-leaf volumes entitled *Auerbach Computer Technology Reports* which are regularly updated. Each report includes a system summary, followed by a detailed survey and tabulated data. The major series is *Standard EDP Reports* but there are also shorter summaries—for instance, *EDP Notebook International*—which contains brief details of equipment available from the major manufacturers throughout the world. Auerbach also covers software in *Software Reports*.

A similar loose-leaf updating service is provided by Hitchcock Publishing Co. of Illinois called *Business Automation Specification Reporting Services*. There are three series: *Computer equipment specifications*, *Peripheral equipment specifications* and *Computer software specifications*.

An extremely compact summary of available equipment is *Computer Characteristics Review* (GML Corporation, Lexington, USA). This claims to list virtually all digital computers and related peripheral devices commercially available, with brief details of each and an indication of price range. It is reissued every 4 months.

From the same source comes *Computer Display Review* (GML Corporation). This is a loose-leaf series covering all types of display equipment for use with computers.

A UK publication which covers similar ground to the above is *Computers in Europe* (Richard Williams, Llandudno, Wales). To date, 4 editions have been published at intervals of 1 or 2 years.

For information on programs which are generally available, the following can be consulted: *Verified software products—a catalogue* (National Computing Centre, Manchester, 1972. Factfinder No. 11) and *Computer programs directory* 1971 (B. R. Faden, ed., CCM Information Corporation, for Association for Computing Machinery, 1971).

There are several publications which are useful for general information on the computer industry. Besides including information on manufacturers' products, most of them cover services such as consultants and service bureaux as well as societies, computer user groups, etc:

Computer Industry Annual (Computerfiles Inc., Concord, Mass.)
Computer Yearbook (American Data Processing Inc., Detroit)
The Computer Directory (IPC Electrical-Electronic Press, London)
The Computer Users' Year Book (Brighton, England)

There are also several trade directories in broader fields, which include computers:

Electrical and Electronics Trade Directory (Benn, London)
Electronic Engineers Master (United Technical Publications, New York)
Electronics Buyers' Guide (McGraw-Hill)
Japan Electronics Buyers' Guide (Dempa Publications, Tokyo)

Appendix A. Acronyms

ACM	Association for Computing Machinery
AEC	Atomic Energy Commission (USA)
AIP	American Institute of Physics, New York
AMCOS	Aldermaston Mechanised Cataloguing and Ordering System
AWRE	Atomic Weapons Research Establishment, Aldermaston
BASIC	Biological Abstracts Subjects In Context (also a computer language)
BLL	British Library Lending division
BNB	*British National Bibliography*
BNL	Brookhaven National Laboratory
CERN	European Organization for Nuclear Research, Geneva
CNRS	Centre National de la Recherche Scientifique, Paris
CODATA	Committee on Data for Science and Technology, ICSU
COM	Computer Output in Microform
COSATI	Committee on Scientific and Technical Information, Washington
COSPAR	Committee on Space Research, ICSU
COSTED	Committee on Science and Technology in Developing Countries, ICSU
CPM	*Current Physics Microfilm*, AIP
CPP	*Current Papers in Physics*, INSPEC
CROSS	Computer Rearrangement of Subject Specialities (*Biological Abstracts*)
DESY	Deutsches Elektronen Synchrotron, Hamburg
ENEA	European Nuclear Energy Agency
ESRO	European Space Research Organization, Paris
FID	Fédération Internationale de Documentation
HEPI	*High Energy Physics Index*, DESY
IAA	*International Aerospace Abstracts*, NASA
IAEA	International Atomic Energy Agency, Vienna
ICSU	International Council of Scientific Unions
IEE	Institution of Electrical Engineers, London
IEEE	Institute of Electrical and Electronics Engineers, New York
INIS	International Nuclear Information Service, IAEA
INSPEC	Information Service in Physics, Electrotechnology and Control, IEE
ISA	Instrument Society of America
IUC	International Union of Crystallography
IUPAC	International Union of Pure and Applied Chemistry
IUPAP	International Union of Pure and Applied Physics
KWIC	Key Word in Context
KWOC	Key Word Out of Context
MARC	Machine-readable Catalogue
NASA	National Aeronautics and Space Administration, Washington
NATO	North Atlantic Treaty Organisation
NBS	National Bureau of Standards, Washington
NRLSI	National Reference Library of Science and Invention (now Science Reference Library)

NSA	*Nuclear Science Abstracts*, US AEC
NTIS	National Technical Information Service, Washington
OECD	Organisation for Economic Co-operation and Development, Paris
OSTI	Office for Scientific and Technical Information (now incorporated into British Library)
PA	*Physics Abstracts*, INSPEC
PASCAL	Programme Appliqué à la Sélection et à la Compilation Automatique de la Littérature, CNRS
PPF	*Preprints in Particles and Fields*
PSI	*Permuterm Subject Index*
RIBA	Royal Institute of British Architects, London
SCI	*Science Citation Index*
SCOPE	Scientific Committee on Problems of the Environment, ICSU
SDI	Selective Dissemination of Information
SLAC	Stanford Linear Accelerator
SPIN	Searchable Physics Information Notices, AIP
STAR	*Scientific, Technical and Aerospace Reports*, NASA
SUN	Symbols, Units, Nomenclature
UDC	Universal Decimal Classification, FID
UKAEA	UK Atomic Energy Authority
Unesco	United Nations Educational, Scientific and Cultural Organisation, Paris
UNISIST	*World Science Information System*, Unesco
USAEC	Atomic Energy Commission, Washington
VINITI	All-union Institute for Scientific and Technical Information, Moscow
ZAED	*Zentralstelle für Atomkernenergie Dokumentation*, Karlsruhe

Appendix B. Publishers

Abrams	Harry N. Abrams, Inc., New York
Academic Press	Academic Press, Inc., New York and London
Addison-Wesley	Addison-Wesley Publishing Co., Reading, Mass.
Akademie	Akademie Verlag, Berlin, DDR
American Elsevier	American Elsevier Publishing Co., New York
Annual Reviews	Annual Reviews, Inc., Palo Alto, California
Arnold	Edward Arnold, London
Athlone	Athlone Press, London
Bell	George Bell and Sons, London
Benjamin	W. A. Benjamin, Inc., Menlo Park, California
Bingley	Clive Bingley, London
Blackie	Blackie and Son, London
Blaisdell	(now) Xerox College Publishing, Lexington, Mass.
Butterworths	Butterworth and Co., London
CCM Information	Crowell, Collier and Macmillan, New York
Chambers	W. and R. Chambers, Edinburgh
Chapman	Geoffrey Chapman, London
Chapman and Hall	Chapman and Hall, London
Chatto and Windus	Chatto and Windus, London
Chemical Publishing	Chemical Publishing Co., New York
Chemical Rubber	Chemical Rubber Co. (CRC), Cleveland, Ohio
Chilton	Chilton Book Co., Philadelphia
Clarendon	Clarendon Press, Oxford
Constable	Constable and Co., London
Consultants	Consultants Bureau, New York
CUP	Cambridge University Press, London
David and Charles	David and Charles, Newton Abbot, UK
Dekker	Marcel Dekker, New York
Dover	Dover Publications, New York
Dunod	Dunod, Paris
Elsevier	Elsevier Publishing Co., Amsterdam, New York (American Elsevier)
English Universities	English Universities Press, London
Faber	Faber and Faber, London
Fitzmatgiz	Foreign-language Scientific and Technical Dictionaries, Moscow
Focal	The Focal Press, London
Foulsham	Foulsham and Co., Slough, UK
Freeman	W. H. Freeman and Co., San Francisco
Gauthier-Villars	Gauthier-Villars, Paris
Gordon and Breach	Gordon and Breach, New York, London and Paris
Harper and Row	Harper and Row, New York
Harrap	G. G. Harrap and Co., London
Hayden	Hayden Book Co., New York
Heinemann	William Heinemann, London

Hermosa	Hermosa Publishers, Alburquerque, New Mexico
Heyden	Heyden and Son, London
Heywood	c/o Butterworth, London
Hilger	Adam Hilger, London (now includes Hilger and Watts)
HMSO	Her Majesty's Stationery Office, London
Holt-Rinehart	Holt, Rinehart and Winston, New York
Hutchinson	Hutchinson and Co., London
Iliffe	Iliffe Books, London
Institute of Physics	Institute of Physics and the Physical Society, London
Interscience	Interscience Publishers, New York (a division of John Wiley)
IPC Science	IPC Science and Technology Press, Guildford, Surrey
Lewis	H. K. Lewis, London
Library Assoc.	Library Association, London
Longmans	Longmans, Green and Co., London
Macdonald	Macdonald and Co., London
McGraw-Hill	McGraw-Hill Book Co., New York
Macmillan	Macmillan and Co., London
	Macmillan Publishing Co., New York (now unconnected)
Masson	Masson et Cie., Paris
Merrill	Charles E. Merrill Publishing Co., Columbus, Ohio
Methuen	Methuen and Co., London
Mills and Boon	Mills and Boon, London
MIT	The MIT Press, Cambridge, Mass.
Mono Book	Mono Book Corporation, Baltimore
Moos	Heinz Moos Verlag, München
Morgan-Grampian	Morgan-Grampian Books, London
Newnes	George Newnes, London
North-Holland	The North-Holland Publishing Co., Amsterdam
Oliver and Boyd	Oliver and Boyd, Edinburgh
Optosonic	Optosonic Press, New York
OUP	Oxford University Press, London and New York
Penguin	Penguin Books, Harmondsworth, UK
Peregrinus	Peter Peregrinus, Stevenage, UK
Pion	Pion Ltd, London
Pitman	Sir Isaac Pitman, London
Plenum	Plenum Publishing Corporation, New York
Prentice-Hall	Prentice-Hall, Englewood Cliffs, New Jersey
Reidel	D. Reidel Publishing Co., Dordrecht, Netherlands
Reinhold	Reinhold Publishing Corp. (now Van Nostrand–Reinhold), Princeton, London
Ronald	Ronald Press Co., New York
Scripta	Scripta Publishing Co., Washington
Spartan	Spartan Books, New York
Spon	E. and F. N. Spon, London
Springer	Springer Verlag, Berlin, Heidelberg and New York
Steinkopff	D. Steinkopff, Darmstadt
Taylor and Francis	Taylor and Francis, London
Trade and Technical	Trade and Technical Press, Morden, Surrey
United Technical	United Technical Publications, New York
University	University Tutorial Press, London
Van Nostrand	D. van Nostrand Co., Princeton and London

Vieweg	Friedrich Vieweg und Sohn, Braunschweig
Weidenfeld	Weidenfeld and Nicolson, London
Wiley	John Wiley and Sons, New York and London
Wilson	The H. W. Wilson Co., New York
Wolters-Noordhoff	Wolters-Noordhoff, Groningen, Netherlands

Index

Author Index*

*Only authors for whom there is a fair amount of descriptive matter in the text are included. Chapters in the present volume are indicated by *italic* page numbers.

Subject and Title Index*

*Indexing has been done selectively: mainly reference and other items of importance. Otherwise books and periodicals have been entered only where there is more than a brief mention in the text. For lists of periodicals, reviews, etc., look under each subject heading, e.g. *Light*. Where coverage is general (whole of physics) there are form entries.